The Irish Woods Since Tudor Times

THE IRISH WOODS SINCE TUDOR TIMES
Distribution and Exploitation

Eileen McCracken MSc PhD

A publication of the Institute of Irish Studies
Queen's University
Belfast

David & Charles
Newton Abbot

ISBN 0 7153 5008 0

COPYRIGHT NOTICE

© EILEEN MCCRACKEN 1971

All rights reserved. No part of this publication may be reproduced, stored in a retrieval system, or transmitted, in any form or by any means, electronic, mechanical, photocopying, recording or otherwise, without the prior permission of David & Charles (Publishers) Limited.

Set in 11 on 13 point Juliana
and printed in Great Britain
by Clarke Doble & Brendon Limited Plymouth
for David & Charles (Publishers) Limited
South Devon House Newton Abbot Devon

For my husband
J. L. McCRACKEN

Contents

		Page
	List of Illustrations	9
	Preface	11
	Dirge of the Munster Forest, 1591	13
Chapter 1	Introduction	15
2	The Distribution of Woodland, 1600–1800	35
3	Timber in Industrial Processes	57
4	The Timber Trade in the Seventeenth Century	97
5	The Timber Trade in the Eighteenth Century	112
6	The Price of Timber and Timberworkers' Wages	122
7	The Timber Merchants	127
8	The Era of Private Planting	135
9	State Planting	142
	Notes and References	153
Appendix 1	Wood acreages in the mid-seventeenth century	161
2	Ships employed in the trade of Ireland, 1753	163
3	Charcoal-burning ironworks, 1600–1800	165
4	Glossary of timber trade terms	169
	Bibliography	171
	Index	181

List of Illustrations

PLATES

	Page
A map of 1580 showing the barony of Vdrone *(Crown copyright; Public Record Office, London MPF 70)*	33
Belfast shipyards, 1812 *(Courtesy of Ulster Museum)*	34
Manor of Great Belan, county Kildare, 1772 *(Courtesy of National Library of Ireland)*	34
An edge-wheel for crushing *(Courtesy of Ulster Folk Museum)*	51
A cot on Lough Erne *(Courtesy of Ulster Folk Museum)*	51
Meeting of the Waters, county Wicklow *(Courtesy of National Library of Ireland)*	52
Killarney Lake from the Kenmare road *(Courtesy of National Library of Ireland)*	52
Richie's Dock, Belfast 1805 *(Courtesy of Ulster Museum)*	85
The Dargle, county Wicklow *(Courtesy of National Library of Ireland)*	85
The timber bridge at Cappoquin, county Waterford *(Dublin Penny Journal, 1834)*	86
Hayesbridge on the river Avonmore at Avondale *(J. Fisher, 'Scenery of Ireland', 1792)*	86
Sitka spruce, county Tyrone *(Courtesy of Ministry of Agriculture, Northern Ireland)*	103
Larch, county Down *(Courtesy of Ministry of Agriculture, Northern Ireland)*	103

	Page
Camus Forest, county Londonderry *(Courtesy of Ministry of Agriculture, Northern Ireland)*	104
Castlewellan Forest Park, county Down *(Courtesy of Northern Ireland Tourist Board)*	104

FIGURES

Map of Killarney (I. Weld: *Illustrations of the scenery of Killarney, 1807*)	41
Wooden house in Drogheda *(Dublin Penny Journal, 1832)*	74
Ancient wooden house in Dublin *(Dublin Penny Journal, 1833)*	76

MAPS

1	Uninhabited areas *(Methuen & Co)*	16
2	Townland names containing the word 'derry' *(Quarterly Journal of Forestry)*	24 and 25
3	Woodlands 1600 *(Irish Historical Studies)*	36
4	Location of well known woods	38
5	Furnaces and forges, 1600–1800	91
6	Dublin timber merchants, 1750–1800	128
7	Woods advertised for sale, 1730–80	133
8	Dublin seedsmen and nurserymen, 1740–1800 *(Quarterly Journal of Forestry)*	139
9	State forests, 1966 *(Society of Irish Foresters)*	144 and 145

Preface

THIS work had as its genesis a thesis on the history of woodland in Ulster under the supervision of Professor Estyn Evans, now Director of the Institute of Irish Studies, Queen's University, Belfast. Over the years it has been expanded to cover the whole of Ireland and to include a study into some aspects of the industries dependent on native and imported timber during the seventeenth and eighteenth centuries.

Dr Evans has always stimulated his students to an awareness of what is only too easily taken for granted and I record with pleasure my thanks to him for arousing an interest in the history of the landscape of my native land.

A great many Irish historians have been generous with help and I am indebted to them. I am most grateful for the willing assistance I received from the staff of the Public Record Offices of Ireland and of Northern Ireland, the National Library of Ireland and the library of HM Customs and Excise, London; from Desmond Clarke, librarian to the Royal Dublin Society and from James Vitty, librarian of the Linenhall Library, Belfast.

I also want to thank George E. McWatters, Group Chairman of Harveys of Bristol, Ltd, R. V. Westrup of Arthur Guinness, Son & Co, Dublin, and W. T. Barter of the Worshipful Company of Coopers, London, who by supplying me with information as to the sizes and weights of different types of staves and sawn timber made it possible to form an estimate of the extent of the Irish timber trade at various times.

As well I thank T. M. McEvoy, Inspector General, Forestry Division, Department of Lands, Dublin, for permission to consult the MS of his thesis on the ecology of woodland flora in Ireland; and Risteard MacGabhann, lecturer in Irish, The New University of Ulster, for help in preparing the list of tree names in Irish.

As to my conclusions, they are my responsibility alone and my personal assessment of the evidence.

The publication of this work has been made possible by a generous grant from the Institute of Irish Studies, the Queen's University, Belfast.

Finally, I wish to record gratitude, which I did not always feel at the time, to my husband, who kept me from straying too far from the woodland rides of fact into the brushwood of historical speculation.

January 1970
Portballintrae,
County Antrim

Dirge of the Munster Forest, 1591

Bring out the hemlock! bring the funeral yew!
The faithful ivy that doth all enfold;
Heap high the rocks, the patient brown earth strew,
And cover them against the numbing cold.
Marshal my retinue of bird and beast,
Wren, titmouse, robin, birds of every hue;
Let none keep back, no, not the very least,
Nor fox, nor deer, nor tiny nibbling crew,
Only bid one of all my forest clan
Keep far from us on this our funeral day.
On the grey wolf I lay my sovereign ban,
The great grey wolf who scrapes the earth away;
Lest, with hooked claw and furious hunger, he
Lay bare my dead for gloating foes to see—
Lay bare my dead, who died, and died for me.

For I must surely die as they have died,
And lo, my doom stands yoked and linked with theirs;
The axe is sharpened to cut down my pride:
I pass, I die, and leave no natural heirs.
Soon shall my sylvan coronals be cast;
My hidden sanctuaries, my secret ways,
Naked must stand to the rebellious blast;

No spring shall quicken what this Autumn slays.
Therefore, while still I keep my russet crown,
I summon all my lieges to the feast.
Hither ye flutterers! black, or pied, or brown;
Hither ye furred ones! Hither every beast!
Only to one of all my forest clan
I cry 'Avaunt! Our mourning revels flee!'
On the grey wolf I lay my sovereign ban,
The great grey wolf with scraping claws, lest he
Lay bare my dead for gloating foes to see—
Lay bare my dead, who died, and died for me.

Hon Emily Lawless

What shall we do for timber?
The last of the wood is down,
....

There's no holly nor hazel nor ash here
But pastures of rock and stone,
The crown of the forest is withered
And the last of its game is gone.

From *Kings, Lords and Commons* by Frank O'Connor
(reproduced by kind permission of
Macmillan & Co Ltd)

CHAPTER 1

Introduction

In 1600 about one-eighth of Ireland was forested; by 1800 the proportion had been reduced to a fiftieth as a result of the commercial exploitation of the Irish woodlands following on the establishment of English control over the whole country.

To say, however, that an eighth of the country, or $12\frac{1}{2}$ per cent, was wooded in 1600 is to give a rather misleading picture. Map 1 shows the virtually unpopulated parts of Ireland. These include land over 5–600ft, which is the upper limit of human settlement and in many cases of tree growth also; the limestone karst lands of the country around Galway bay and the Shannon-Suck interfluve; and the lowland parts of the country which are covered with either blanket or raised bog. In fact considerably more than $12\frac{1}{2}$ per cent of the land suitable for agriculture was then forested.

There are few statistics relating to Irish woodland before the middle of the seventeenth century, when surveys of certain parts of most counties were carried out, but even these do not provide the detailed information that can be produced for various contemporary English forests. It is known, to take a few examples, that in 1608 there were 124,000 trees in the New

Map 1. Uninhabited areas, bog and upland (from: *Ireland, its physical, historical, social and economic geography* by T. W. Freeman)

Forest and 23,400 trees in Sherwood Forest, and that in 1633 the Forest of Dean contained 165,000 trees.[1] The most that can be said of Irish woods is that some were of greater extent and thicker than others: Glenconkeyne on Lough Neagh was denser than Kirlwarlin in the Lagan valley and Cork had a greater area of woodland than Donegal. It is also a fact that the deciduous woods generally did not extend above 500ft though in Killarney and Wicklow they pushed up to nearly 1,000ft.

The foundations of Ireland's woods were laid during the amelioration of climate which followed the final Ice Age. Species of birch, willow, pine, hazel, elm, oak, ash, yew, mountain ash, alder, juniper, bird cherry, whitebeam, and holly were finally established in Ireland after the major climatic fluctuations of the post-glacial age had given place to the type of climate which, with minor changes, has prevailed since about 500 BC. Certain large deciduous trees—horse chestnut, sweet chestnut, sycamore, poplar, beech and lime are not indigenous and were introduced in recent centuries. Scots pine, which at two separate post-glacial periods had formed extensive forests, may have survived in isolated pockets, but was absent from nearly the whole of Ireland until reintroduced about the middle of the seventeenth century by Cromwellian soldiers. The presence of the capercaille—the Cock of the woods—in Ireland in 1684[2] may point to the existence of pine woods, but the evidence is not conclusive as the capercaille has been found on occasions in deciduous woods.

Elm, like pine, was once widely distributed but subsequently almost completely disappeared. Its virtual elimination by the seventh century AD has been ascribed partly to disease and partly to the practice of feeding tethered cattle on its foliage. It is clear, however, from the evidence presented by A. C. Lucas that in the early Christian era elm was well known and widely distributed in Ireland[3]. No contemporary reference is found to elm in Ireland until the eighteenth century, apart from those planted in the occasional garden, and it was necessary to import it for mounting artillery at the end of the sixteenth century.

Twice at least appeals were sent to England for elm: in 1590 Lord Carew wrote to Lord Burleigh that the artillery in Galway and Limerick lay upon the ground unmounted, 'which for want of elm planks in Ireland cannot be redressed'; and in 1624 a request was made for the dispatch of 130 tons of elm timber and plank for use in Dublin and Limerick[4].

A Restoration traveller in Ireland remarked that elm and fir were rarely seen growing in the kingdom[5]. It is possible that wych elm may have survived in the Glens of Antrim, though a list of trees in the Glens in 1740 does not include it[6]. Advertisements for the sale of elm and pine first appear about 1740, but of the 200 odd advertisements for the sale of timber which appear in Faulkner's *Dublin Journal* between 1731 and 1763 only five list elm. This is in line with the number of times other introduced trees were advertised for sale—nine advertisements for pine, three for beech, five for sycamore and four for poplar. After the 1760s references to elm become more frequent: between 1764 and 1780 fourteen advertisements for the sale of elm are found, half of them describing the trees as large or fully grown and eight advertising elm mixed with other introduced species.

As in England elm was in demand for water-pipes during the eighteenth century; it was not until 1802 that metal pipes were used in Ireland. Dublin Corporation advertised regularly for elm, specifying that the pieces were to be over 10ft long and the diameter at the smaller end to lie between 6 and 22in. The elms along the Dublin Royal Canal were originally planted to provide timber for such pipes.

The historical records for the seventeenth and eighteenth centuries provide no clues as to whether the arbutus and the pedunculate oak are native or introduced. Threlkeld (1726) recorded the arbutus as growing round Lough Lean, in Kerry, and Glengarriff, Bantry Bay, where it is said to have been first noticed in 1640. A little later Sir William Petty remarked that 'it groweth in great numbers and beauty' in Kerry[7]. Arbutus was also found round Lough Gill—where Yeats' Lake Isle of Innisfree lies—near Sligo.

Juniper is mentioned by Threlkeld and also by K'Eogh (1735), who described it growing amid rocks near Ardraghen in Galway and in the Burren in Clare, but no seventeenth-century references to its presence have been found.

Hazel was once widespread to a degree that is hard to imagine today, and was often found in association with oak. By the eighteenth century it was much less common. Less than a dozen advertisements for the sale of hazel appear in contemporary newspapers and only one, for about 20 acres at Clonard in Meath, was for hazel alone; in all other cases it was growing with oak. Corroborative evidence for the decline in the distribution of hazel comes from a pollen analysis made at Littleton bog, near Thurles in Tipperary. Here, until just before the end of the seventeenth century, hazel pollen formed 30–40 per cent of all pollen grains but after 1700 it formed less than 5 per cent[8].

Although yew is native to Ireland few references to it are found in the seventeenth century. It seems to have been fairly widespread and there was a yew wood on the southern slopes of Crookdooish in Londonderry. On the whole it did not form large stands but was fairly common as an occasional tree. The Irish yew, *Taxus fastigate*, is reputed to have originated near Florence Court, Fermanagh, in the eighteenth century.

In medieval Ireland yew was often associated with sacred sites and churches. Newry, county Down, is the anglicised form of the Irish *iubhar* (yew) and according to tradition it took its name from the tree St Patrick himself planted there. The original tree was burned, together with the monastery, in 1162. The sacred yew at Clonmacnoise was destroyed by lightning in 1149. St Columcille when in exile remembered the yew which grew in front of his church at Derry and lamented[9]:

> This is the Yew of the Saints
> Where they used to come with me together.

> Ten hundred angels were there,
> Above our heads, side close to side.
>
> Dear to me is that Yew tree;
> Would that I were set in its place there!
> When I entered into the Black Church.

Giraldus Cambrensis, a twelfth-century visitor to Ireland, wrote, 'Yews, with their bitter sap, are more frequently to be found in this country than in any other place I have visited; but you will see them principally in old cemeteries and sacred places where they were planted in ancient times by the hands of holy men, to give them what ornament and beauty they could'.

Ireland in the early seventeenth century lacked the variety of trees that it possessed in the eighteenth: no cedar, pine, spruce, or cypress carried their green through the winter months; holly, yew and juniper, none of which were extremely common, alone represented the evergreens. Of the broad-leaved trees, oak was by far the most widespread. Ash was less common except where limestone was sufficiently near the surface to give a calcareous soil. Alder was abundant where the water-table was high, and hazel was frequently associated with oak. The only other common species was birch.

The present-day pattern of hawthorn hedgerows interspersed at regular intervals with a single tree was unknown. These hedges, with their individual ash, elm, sycamore, chestnut, or oak were an innovation of the late seventeenth century and did not spread far beyond the Pale until the mid-eighteenth century.

Commercial exploitation of Irish woodland by English settlers was carried out during the seventeenth and eighteenth centuries, but a steady reduction in the area and denseness had been going on over the previous centuries. The woods had been depleted by the demands of a virtually self-sufficient rural society: timber was needed for houses, implements, and fuel; land was needed for the growing of oats—the potato had not yet arrived on the scene—and this could involve the clearing of woodland. Cattle, sheep, and goats produced changes in the forest: goats are nimble and destroy any vegetation within their reach and cattle browse on foliage and saplings. Whether or not the rabbit (a Norman introduction) was sufficiently numerous to have had a deleterious effect on woodland can only be guessed, but as Ireland at the end of the sixteenth century abounded in its natural enemies—wolves, foxes, eagles and wild cats—its numbers were probably kept effectively in check.

It has been suggested that fires destroyed parts of the woods but this does not appear very likely. The native Irish woods were all of deciduous trees and hardwoods are difficult to burn; nowadays they are used as fire protection belts in the inflammable conifer woods. The presence of blackened stumps in the bogs may have given rise to this belief originally. A tradition current in parts of Tyrone at the end of the eighteenth century was that fires in the local bog timber were caused by the wind blowing from the south-west for three years, 'and with the dint of the trees rubbing against one another they caught fire on the tops of the hills'[10]. Threlkeld declared that he could not assess the validity of another belief, in some parts of Ireland, that bog timber was the remains of woods planted by the Danes and 'after their expulsion cut down and left to be buried in the earth by the natives to extinguish the Badge of their Servitude'.

Because trees are the biggest of all living flora their removal makes very noticeable changes in a landscape, and it is not only in Ireland that old people remember, perhaps with nostalgia, the time when the countryside was less bare. Traditions of vanished woodland were first recorded about the end of the eighteenth century. To emphasise how extensive the woods had once been it was said that a man or a bird, or an animal—usually a squirrel, wild cat, or pine marten—could have walked from one place to another on the tops of the trees. A Cheshire version runs,

>From Blacon Point to Hillree
A squirrel could leap from tree to tree.

Such traditions, though not in the form of a rhyming couplet, were current in the Glens of Antrim, the north and east shores of Lough Neagh, the Bann valley, the Newry corridor, Lough Erne, Leitrim, Galway, Clare, south Cork, Kilkenny, Tipperary, and Wexford.

Even earlier traditions appear in certain townland names that embody the names of trees. (Townlands are the smallest divisions of land in Ireland and were probably the land occupied by a family. They vary in size, in poor mountainous land tending to be much larger than in agricultural land. There are about twenty to a parish.) Townland names were probably more or less fixed about the eighth century and as clearance of woodland was a steady, if possibly slow, process during the following centuries, a townland name in itself cannot be taken as indicating that the wood still remained in the area until Tudor times. It must also be remembered that the original Irish word may have become corrupted by the passage of time or by the anglicisation that has taken place of many Irish words, so that the original meaning is not clear beyond all reasonable doubt. For example, the Irish *coill*, a wood, and *cill*, a church, have both frequently become *kil*, so that Kilmore can be interpreted as meaning either the Big Wood or the Big Church. Another Irish word for a wood, *ross* or *ros*, generally denotes a wood in the southern half of the country but a peninsula in the northern part: Roscommon is Saint Comon's Wood but the Rosses in Donegal are peninsulas.

The following list gives the Irish and anglicised form of the names of native trees[11].

Species	Irish name	Anglicised form
birch	beith	behy, beha, beagh, behagh, veha, vehy
holly	cuileann	cullen, cullion
oak	dair	dar, der, dara, darra, darragh
blackthorn	draighean	dreen, drain, drin
sallow	saileach	sillagh, sallagh, sill
alder	fearn	far, fern, farnagh, ferney, farnane, farnoge, navarn, navern, navarna

Species	Irish name	Anglicised form
ash	fuinse	funcheon, funshin, funshinagh funchoge
yew	iúr	ure
elm	leamhán	levan, levane, livaun, laune, lamph
whitethorn	sceach	skeagh, skehy, skey, ske, skeha, skew
elder	trom	trim, trom, trum
hazel	coll	coll, col, cole, cull, coyle, kyle, quill

Collective terms

a wood	ros	ross, rus, rush
	fiodh	fee, fi, feigh, feth, fith, fid
	coill	kil, kyle, cuill, cullia
a yew wood	eochaill	aughal, youghal
a river thicket	gaortha	gearha
a grove	garrán	garran, garrane, garraun, garn
an elm wood	leamhchoill	laughil, laghil, laghile, loghill, loughill, lamfield, longfield
an oak wood	diarbhre	darrery, dorrery, darraragh, derravara
an oak grove	daire or doire	derry, derri, der

Oak was the most common species and the anglicised form is included in many townland names. There are about 62,200 townland names in Ireland, and about 1,600 contain derry in one form or another, either as a prefix or suffix, and in some cases the name stands alone. Map 2 shows the distribution of these various forms of derry.

It can be seen from the map that the name is not evenly distributed over Ireland but tends to be found in three regions: (1) the belt of country from the southern shores of Lough Neagh to Westport, (2) central Ireland to the Shannon estuary, and (3) the southern peninsulas of Cork and Kerry.

About a third of all the occurrences of derry are found in the counties of Monaghan, Fermanagh, Armagh, Cavan, Leitrim, Longford, and Roscommon, where the frequency is 24·4, 21, 20, 17·4, 16·8, 13·5 and 13·2 respectively per 100,000 acres.

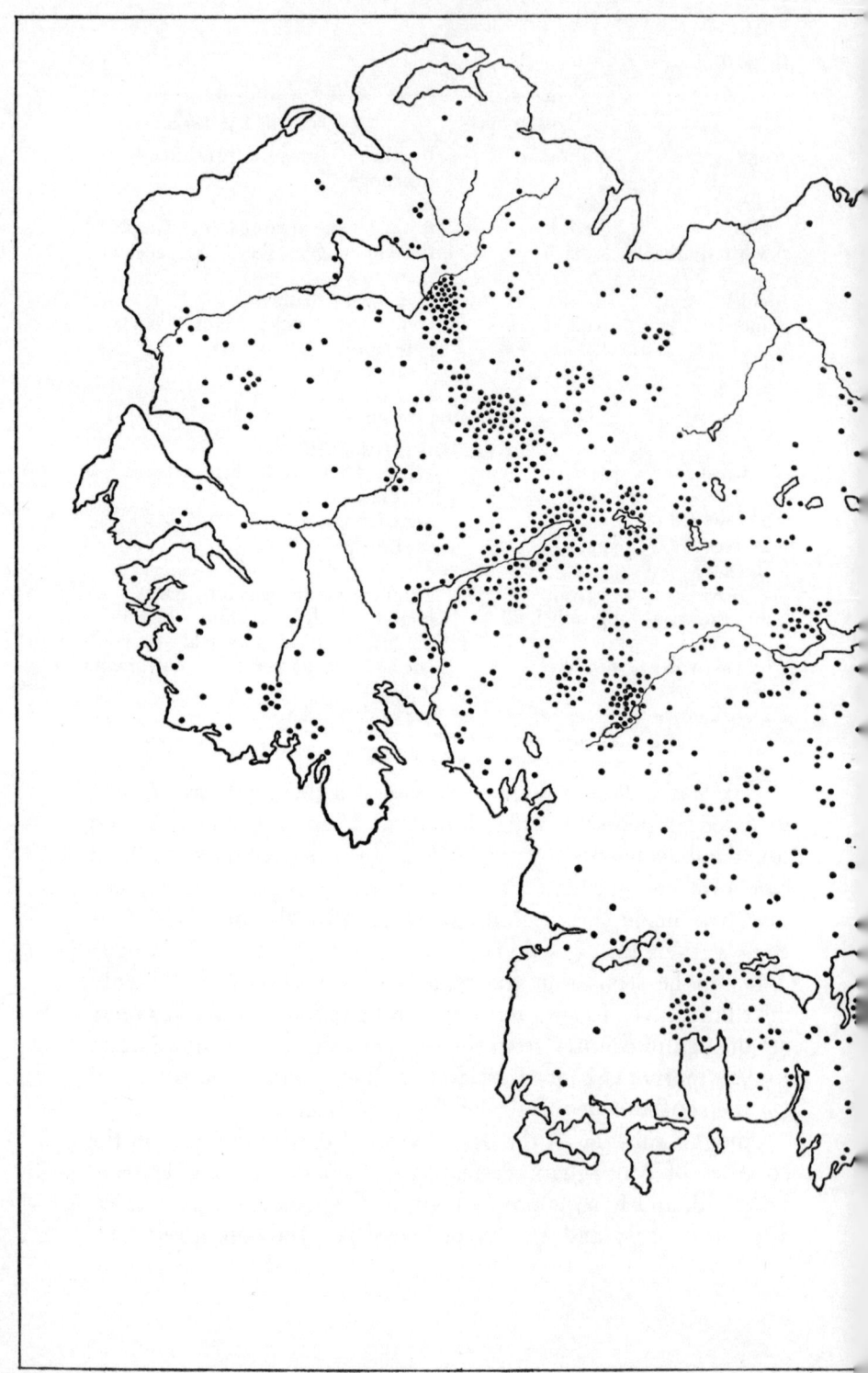

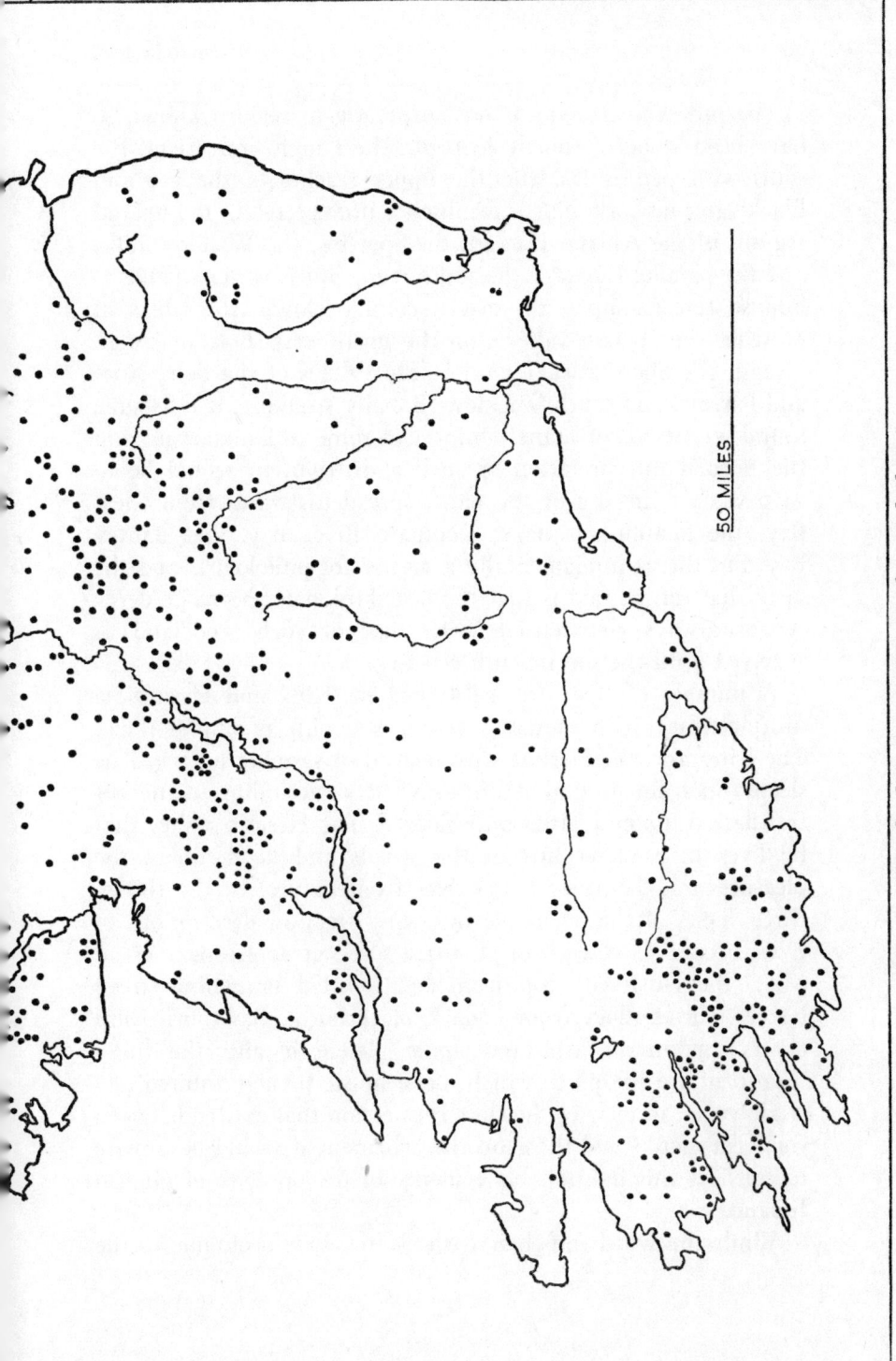

Map 2. Townland names containing the word 'derry'

The presence of derry is not surprising in regions known to have been wooded—north Armagh, the Lough Erne basin, the south-west peninsulas, and the upper reaches of the Lee and Blackwater in Cork. Nor is its absence unexpected in the upland regions of the Antrim Plateau, the Sperrins, the Wicklow hills, and the parallel ridges of the south-west. But it is remarkable to find so few examples in eastern county Down, the Glens of Antrim, the Sperrin valleys, on the north-west shore of Lough Neagh, the Sligo coast, Limerick, the valleys of the Suir, Nore and Barrow, and south Wicklow. Broadly speaking, it is a name found westwards of a line joining Coleraine to Dungarvan, and the normal interpretation of such a distribution would be to suggest that the use of the name spread eastwards from Clew Bay, the Shannon estuary, Kenmare River Bay, and Bantry Bay. But this is fundamentally a matter for philologists and the most that can be said is that the distribution of the name derry is not always co-extensive with that of such woodland as survived until the seventeenth century.

A number of trees are mentioned in 'King and Hermit', a ninth-century Irish monastic poem, and pine is among them. The King asks the Hermit why instead of sleeping on a bed he sleeps 'upon the ground of a fir-grove' [12], or according to another translation 'upon a pitch pine floor'[13]. The Hermit replies that he lives in a secret hut in the woods and he describes the pleasures and bounties he receives from nature. In the thirty-three verses of the poem the following trees, or descriptions of trees, occur: 'a yew-green yew tree', 'the great greenery of an oak', 'the clustered crop from small-nutted branching green hazels', 'black sloes from a dark blackthorn', hawthorn, bird cherry, mountain ash, and apple. There is also the line: 'Beautiful are the pines which make music for me unhired'.

However, in view of the close connection that existed between the Irish monks and those on the continent it would be unwise to consider this line reliable evidence of the presence of pine in Ireland.

While the woods on their Irish estates were profitable to the

Introduction

new settlers they had their drawbacks. They offered shelter to dispossessed Irish and they harboured wolves.

There is an oft-repeated tradition that large tracts of woodland were cut in the early seventeenth century to destroy the cover for those displaced Irish (the woodkernes) who had retreated to the woods and who 'stood upon their keeping' there. But the evidence is meagre and in any case it does not seem a feasible proposition to have axecut large woods in a short time. True, during the previous century, and earlier, it was sometimes suggested that large areas of forest should be cut. Richard II seems to have been the first to put forward such ambitious schemes. In 1399 when McMorough, king of Ireland, lay in the woods west of Kilkenny with 3,000 men, Richard ordered the mobilisation of 2,500 natives to cut down the wood and burn the trees[14]. However, a contemporary remarked, 'In my opinion it was impossible to be effected while the leaves were upon the trees, but after that time, when the trees were bare, then to burn the woods would be the next best means to do service upon him'.

Sir Warham St Leger wrote to Lord Burleigh in 1579 during the Desmond revolt that he proposed to employ 4,000 English soldiers, besides those already in the field, to protect labourers set to cut and burn the woods of Agherlow, Drumfin, Glenmore and Glenflesk in Munster, but there is no evidence that his scheme was ever attempted and in any case these woods were still standing well into the following century. The idea was, however, still being played with in 1585 when Perrott suggested that the Munster woods should be cut 'to deprive the rebels of their places of succour', and he alleged that there were Englishmen willing to undertake the task if the government would lend them £5,000 for three years[15].

There were tracks, or passes, through the woods which had to be kept clear. An Act of 1612 directed that all trees and bushes growing on the highways and passes had to be cleared by their owners. It was pointed out that the 'passages throughout the woods of this kingdom' were, in many places, difficult and dangerous to travel through. Even by the middle of the seventeenth century in well settled areas the passes had to be kept open—to about 60ft on either side of the road.

There is no doubt that the settlers of the early seventeenth century, and especially the military, looked on woods and wood-covered bog as inconvenient refuges for woodkernes. 'The woods and bogs are a great hinderance to us and help to the rebels', wrote an Elizabethan in 1601. 'Much good could be done by Irish churls felling, dressing and burning the trees in heaps. This could be done whilst leaving sufficient timber for the use of the country, if a tree is left standing every twenty yards. Many people think it would have been well if Ireland had been turned into a seapool than have so charged Her Majesty[16].'

Blennerhasset in 1610 described the wolf and the woodkerne as the most serious dangers to the Ulster colonists and recommended periodic manhunts to track down the human wolves to their lairs. 'No doubt', he observed, 'it will be a pleasant hunt and much prey will fall to the followers.'[17]

After 1646 the name woodkerne was generally replaced by tory, from the Irish tóir—search: a tory was one who was searched out. One finds it used in a proclamation of 1656, 'No mercy will be shown to tories', and a later proclamation of 1660 offered rewards varying from £3 to £10 for the betrayal of 'a tory or woodkerne'. The name was well established in popular usage by 1670: a nursery rhyme of that date runs[18],

> Ho brother Teig, what is your story?
> I went to a wood and shot a tory;
> Was it the same, or was it his brother?
> I hunted him in, and I hunted him out,
> Three times through the bog out and about,
> Till out of the bush I spied his head,
> So I levelled my gun and shot him dead.

Tory-hunting was practised at the end of the century. Sir William Stewart of Newtownstewart, co Tyrone, could write in 1683, 'The gentlemen of the country have been so hearty in that chase that of thirteen in the county where I live, in November the last was killed two days before I left home'[19]. Wolves and woodkernes were commonly bracketed together, for they inhabited common ground and represented a common threat to the settlers; and there were rewards for the destruction of wolves as well as of woodkernes.

A grant was made to Henry Tuttesham in 1614 to go to Ireland to destroy wolves. He had to provide his own men, dogs, and traps, and to keep four men and twelve pairs of hounds in every county for the next seven years. In return he could claim four nobles (about £3) for every wolf head. During the Commonwealth a special effort was made to exterminate them. Rewards of between 10s and £6 according to the age and sex of a wolf were offered and also 5s for a fox. Galway, Mayo, Sligo, and part of Leitrim, between them claimed £243 in 1655 and in 1656 the amount paid out for the whole of Ireland was £3,874. In Leitrim during the 1680s a wolf-killer received 2d per hearth in the parish for every wolf he killed[20].

As well as offering rewards for killing wolves, at least two professional wolf-hunters were licensed during the Commonwealth. In 1652 a Richard Toole and a servant were given permission to move about the country with two fowling pieces seeking wolves. The next year a drive was made against wolves in the neighbourhood of Dublin. A Captain Edward Piers was leased land for five years in Dunboyne, county Meath, on terms which included keeping up a hunting establishment for the hunting of wolves and foxes. He was to maintain three wolf hounds, two English mastiffs, and a pack of hounds of sixteen couples (three of them to hunt wolves only), a working huntsman, two men and a boy. An orderly hunt was to take place at least three times a month and a part of the establishment was to be kept at Dunboyne and a part in Dublin. As security against the performance of his duties Piers had to deposit £100 annually in addition to his rent and he was to destroy at least fourteen wolves and sixty foxes in five years[21].

Despite the efforts made during the Commonwealth to

exterminate them, wolves lingered in remote areas well into the eighteenth century. In England they were extinct before 1500, in Scotland by about 1743, but in Ireland the last wolf died about 1770. They disappeared at various times in different parts of the country. Collin mountain behind Belfast was infested with them in the 1660s, as was the eastern side of Lough Neagh. In the latter area some leases, which probably date from the early eighteenth century, bound tenants to kill a certain number during their tenancy. At Waringstown, county Down, they disappeared about the turn of the century, in Cork and Kerry they became extinct about 1710, in Nappan on the Antrim plateau in 1712, but in Glenconkeyne they lingered until the 1760s. They survived longest in the south-west of Limerick and near Feakle in Clare[22].

Before the time of the Commonwealth Irish wolf dogs, as they were then called, were sent both to England and the continent in pairs either as gifts or as sales, but in 1652 their export was prohibited, no doubt to retain as many as possible in the country to aid the Commonwealth drive against wolves. These large light-coloured dogs became extinct about the end of the eighteenth century, the present-day Irish Wolfhound being evolved during the nineteenth century from crossings between Scotch deerhounds, Great Danes, mastiffs and borzois.

Whether or not the new landlords were able to exploit the timber on their estates commercially, if they wished to develop the estate the wood had to be cleared. There are many examples of seventeenth-century leases which require a tenant to clear so much timber annually. An Armagh farmer writing in 1750 declared that he would have had a hard bargain of his farm if a memorable storm had not levelled much of an extensive wood on his holding.

There were cases where the landlord wished to clear scrub only and preserve the good timber. An example of such estate management was found on the Brownlow estate in Lurgan, county Armagh, in the 1670s[23]. To facilitate the clearing of underwood and scrub Arthur Brownlow sold the rights to Edward Hall, who owned an ironworks in Magheralin. By the terms of the agreement Hall paid 4d a cord for a maximum of 700 cords a year. Brownlow's tenants were to be permitted to gather wood for domestic uses and building and Hall was not to take saplings or any wood except that fit for firewood on pain of forfeiture and a fine of six times the wood's value. The cords were not to be converted into charcoal until Brownlow's wood-ranger had inspected and measured them, and Hall was to work systematically from townland to townland until all the underwood was exhausted.

The leases for tenants on the same estate contained clauses binding the tenants to plant orchards, and set ditches with hawthorn and saplings of ash, oak, hazel or sycamore 30ft apart. They were allotted wood for domestic use but were not to burn peat for fuel while underwood was still available.

When timber became scarcer towards the end of the seventeenth century the right of the tenant to cut wood tended to become more restricted. An early lease of this type comes from the Rawdon estate on the east side of Lough Neagh. A lease of 1653 empowered the tenant to cut timber only on the express permission of the landlord. The Irish Society inserted various clauses into their leases early in the eighteenth century limiting the right of the tenant to timber. After August 1715 the Mercers' Company bound their tenants to preserve young trees and the following month the Grocers', Fishmongers', Goldsmiths', and Skinners' Companies examined the leases of their tenants with the object of clarifying the position of timber on their estates. The Irish Society decided in 1720 that the timber allowance to tenants did not include the lop and top of timber felled by the Society[24].

To a tree-conscious landlord interested in improving his woods the allowance of wood to tenants could be an irritant. The land agent on the Abercorn estate in county Tyrone complained of the inconvenience of granting rights to tenants: he wrote in 1744 that they prevented the proper growth of woods which although of little value at the moment could be greatly improved by proper care[25].

A series of parliamentary acts between 1689 and 1791 tried either to conserve existing timber or to encourage or enforce planting. The extent to which these acts were enforced depended not only on the personal whim of the landlord but also on how seriously the grand juries of counties performed their duty of enforcing them. Broadly one may say that after 1721 tenants on estates were entitled by statute to a third of all the trees they planted, a half after 1731, and all of them after 1765, but these provisions related only to planted timber. However, after 1767 persons cutting down trees without the owner's consent had to pay treble their value and life tenants became impeachable for waste of timber in 1783. The last of the Acts on the subject of tenant and timber was passed in 1791 and laid down that no person holding lands by lease could fell timber unless by covenant in the lease or by consent of the owner.

(See pages 154–5 for notes)

Page 33: This map, made about 1580, shows the north-west of county Carlow lying between the Kilkenny border on the west and the river Burren, a tributary of the Slaney, on the east. Despite reasonably close settlement quite extensive tracts of oak wood remained on the lowlands. The note at the bottom left-hand corner reads: 'In this barony of Vdrone, alias Idrone, hath been one hundreth villages or townships, some with churches and castles more than there now is which hath been destroyed by the uncivilised order of the nations or kindreds late inhabiting and ruling the same barony'.

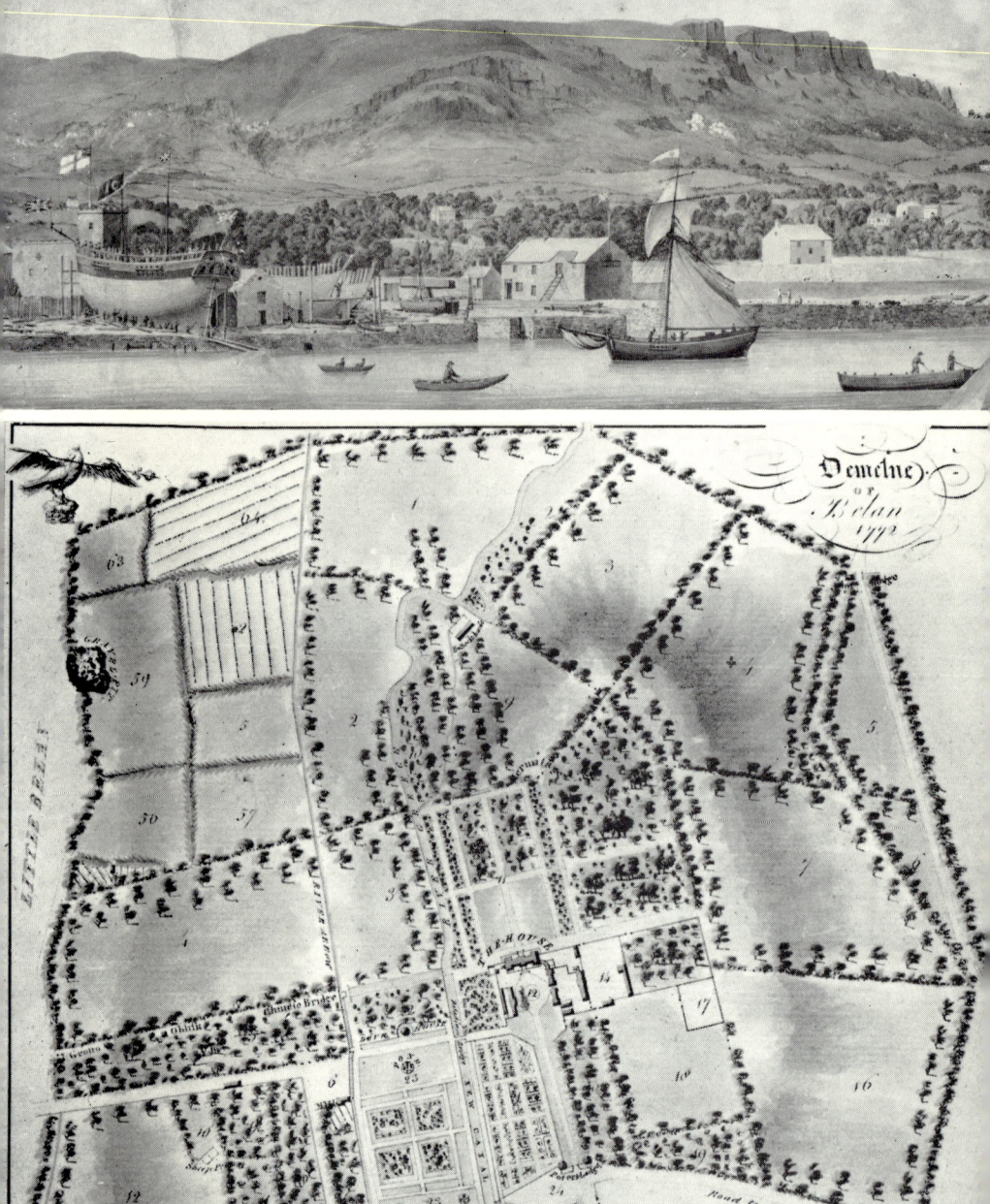

Page 34: (above) Belfast shipyards, 1812. This painting shows a large wooden vessel being launched into the Lagan estuary and another one in course of construction. The remains of the oak wood that had once extended up the Lagan valley lie along the lower slopes of Cave Hill; *(below)* Manor of Great Belan, county Kildare, 1772. This map shows planting carried out on the estate of the Earl of Aldborough near Castledermot about the beginning of the eighteenth century. The grounds are laid out in the formal pattern characteristic of the period. The trees are planted in blocks of one species—for example, an elm grove (19), an oak grove (5) and a fir grove, probably Scots pine (8).

CHAPTER 2

The Distribution of Woodland, 1600-1800

The distribution of woodland in Ireland at the beginning of the seventeenth century is shown on map 3. The largest and densest areas of woodland lay to the north-west of Lough Neagh, in the Erne basin, along the Shannon, in the river valleys of the west and south, and on the eastern slopes of the Wicklow and Wexford hills. Smaller but significant areas of wood existed in eastern county Down, in the Glens of Antrim, in the Sperrin valleys, on the western coast of Lough Swilly, on the western coast of Donegal, and in north Sligo and south Galway. Woodland was much less prevalent in the central part of Ireland, which included the long-settled Pale (the counties of Dublin, Kildare, Meath and Louth) and the boggy land of the central plain[1].

The location of woods having distinctive names is shown on map 4. In county Antrim the Glens, which 'can show you many a sight', lay along the coast, and in the extreme south-west of the county, Killultagh, the Wood of Ulster, formed a triangle lying between Lough Neagh, Portmore Lough and the Maze.

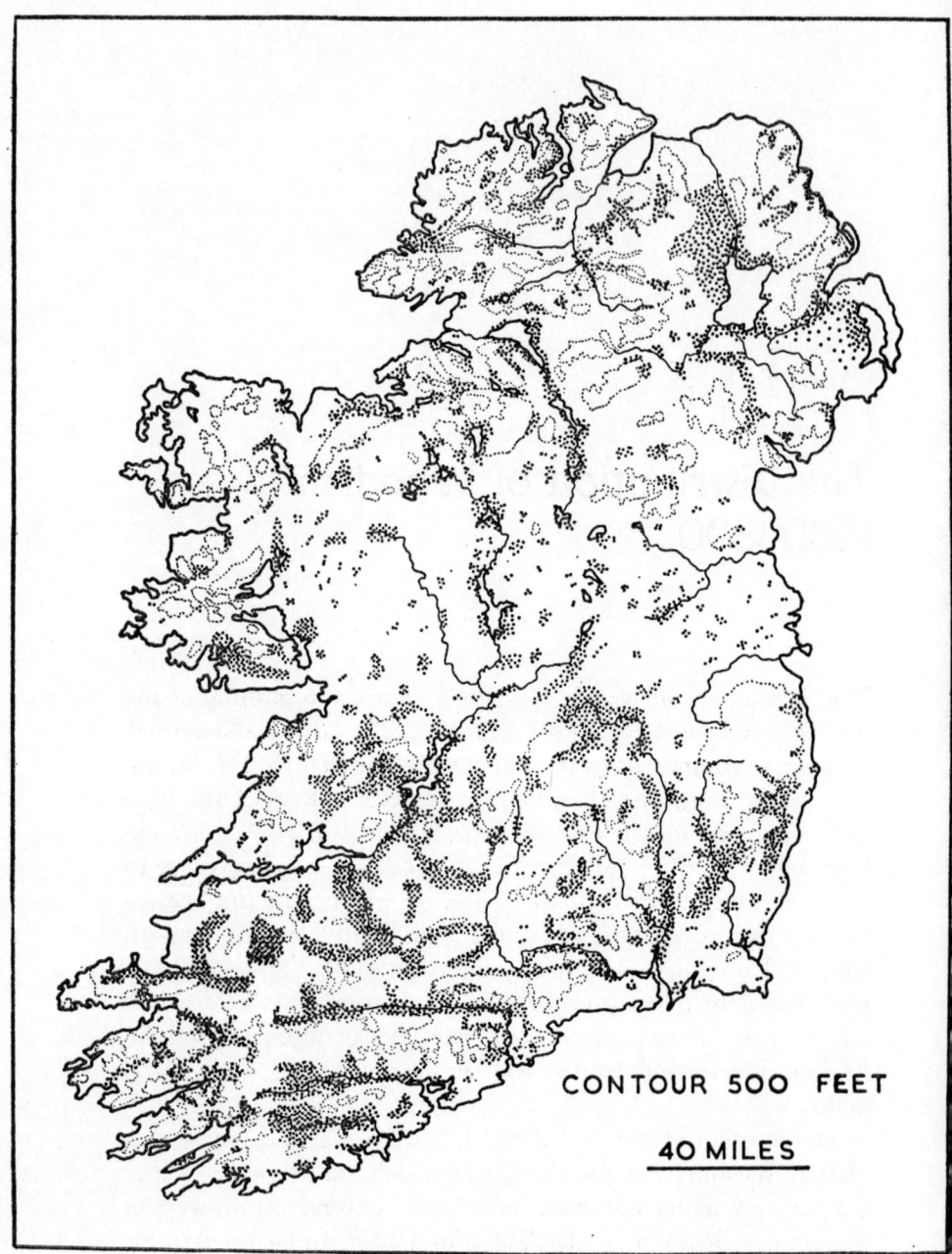

Map 3. Woodlands 1600

On the south side of the River Lagan, between Strangford Lough and Lough Neagh, were the three wooded areas called the Dufferin, McCartan's Country and Kilwarlin. The Dufferin lay for the most part below 200ft, while McCartan's Country covered the higher ground towards Slieve Croob and Ballynahinch. Kilwarlin was the territory south of the Lagan. In south Down, north-west of Newry and surrounding Slieve Gullion, was the Fews.

The woods and wooded fens on the southern shore of Lough Neagh were designated by the district names of Clancan, Clanawle, Clanbrassel and O'Neilland: between the Blackwater and Bann rivers near the lough shore lay Clancan; to the south Clanawle stretched from the Blackwater to Armagh, and O'Neilland from Armagh to Portadown; and eastwards from Portadown to the Armagh-Down border was Clanbrassel. On the north-west shoulder of Lough Neagh were Mountreivelin, Killetra, and Glenconkeyne, the latter merging with the woods of the lower Bann valley.

In Connaught, the Leguy woods in Sligo lay on the north slopes of the Ox mountains and county Roscommon contained the Fasach-Coille near Lough Allen, the Coill Conchobhair south of Boyle, and the Feadha or Fews west of Lough Ree. The Wood of Suidain lay west of Lough Derg on the Clare-Galway border.

In Munster, where the extent of forest was almost legendary in Tudor times, Glanekinty was north of Tralee and the Killarney woods grew around Lough Leane. The Great Wood of Kilmore occupied part of the Limerick-Cork border, and Aherlow lay in a Tipperary valley at the foot of the Galty hills. The Slieve Groot Wood grew at the western end of the Knockmealdown hills.

Three names need to be mentioned in Leinster: the Great Wood of Kilconish in the lower Suir valley east of the Commeragh mountains, the Wood of Coillaughtim on the Slaney north of Enniscorthy, and Shillelagh in south Wicklow.

These areas of woodland, which at the beginning of the

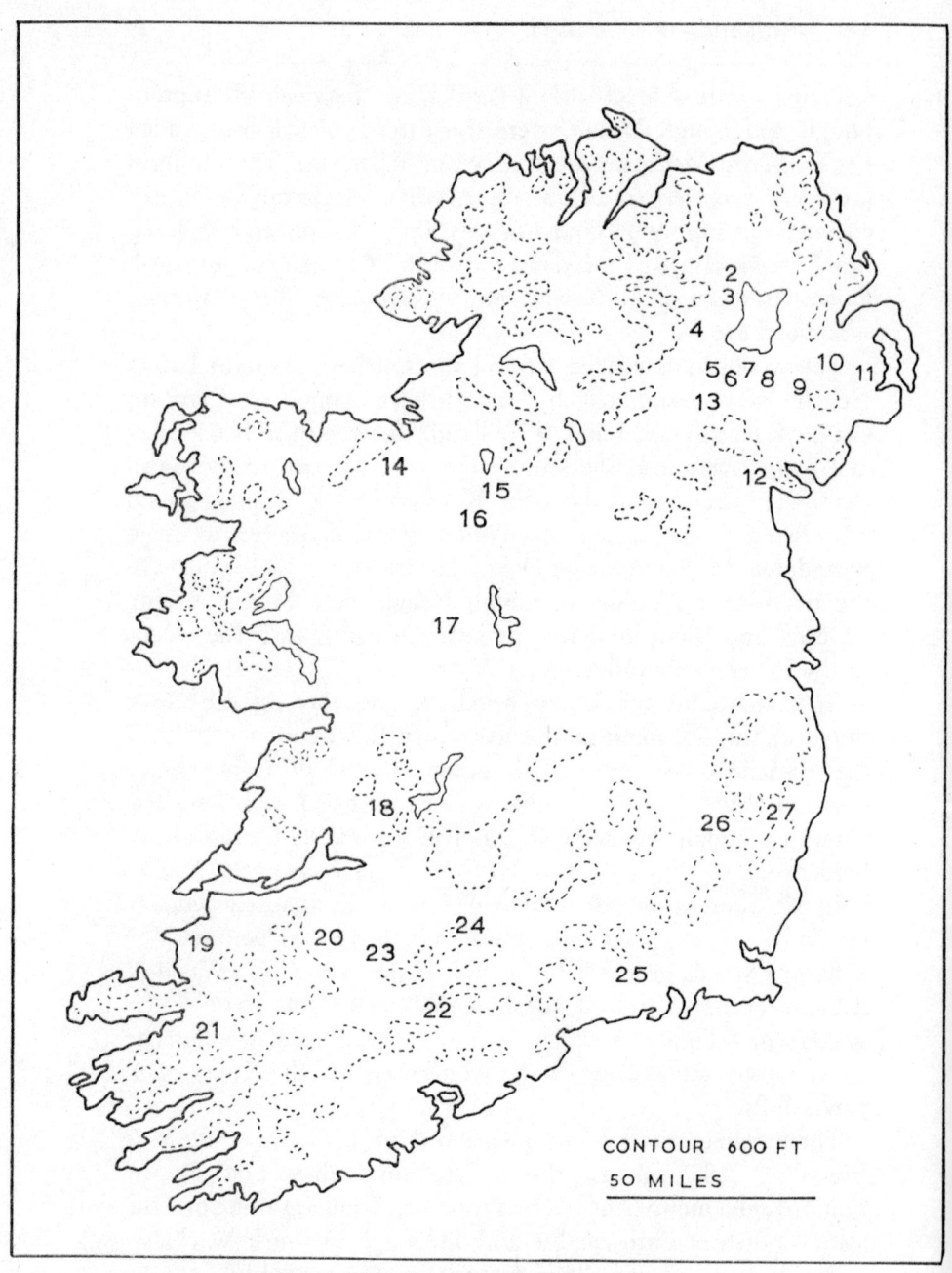

Map 4. Location of well known woods

The Distribution of Woodland, 1600–1800

seventeenth century covered about an eighth of the whole country, were progressively diminished until by the beginning of the nineteenth century probably only 2 per cent of the land was forested. The woods then remaining were largely on the outlying areas—the Glens of Antrim, the remote north-west of Donegal, Killarney, and the eastern slopes of the Wicklow hills.

The end of the eighteenth century, too, saw the passing of the old landscape for ever. The dominant trees had been oak or ash, with a sprinkling of hazel, holly, alder and willow—deciduous hardwoods whose appearance changed with the seasons. The nineteenth century was to see the rise of the conifers and the dilution of native hardwoods by elm and beech, sycamore, chestnut, and lime.

Let us now examine Irish woodland in the seventeenth century in greater detail.

On the Antrim coast the four Glens of Glenarm, Carnlough, Glenariff, and Cushendun carried oak on the wet valley bottoms and on the lower slopes, while on the higher and drier ground the oak gave place to hazel, holly, ash, and alder. This timber was used for burning chalk for fertiliser during the early eighteenth century and it was told in the early nineteenth century that once a man could have crosssed the valley of any of the Glens on the tops of the trees and that a bird could have

1	The Glens	15	The Fasach-Coille
2	Glenconkeyne	16	The Coill Conchobhair
3	Killetra	17	The Feadha or Faes
4	Mountreivelin	18	The wood of Suidain
5	Clancan	19	Glanekinty
6	O'Neiland	20	Clonish wood
7	Clanbrassel	21	Killarney
8	Killultagh	22	Slieve Groot
9	Kilwarlin	23	Great wood of Kilmore
10	McCartan's Country	24	Aherlow
11	The Dufferin	25	Great wood of Kilconish
12	The Fews	26	Coillaughtim
13	Clanawle	27	Shillelagh
14	The Leguy woods		

hopped the length of a Glen on the branches[2]. Hazel was particularly widespread between the Antrim Plateau and the sea on the southern part of the Antrim coast and some ash, birch, and alder was also present together with heavy oak forest in the valley of the Glynn river.

One of the biggest areas of woodland in the country and possibly the densest, was the oak forests of Mountreivelin, Killetra, and Glenconkeyne, which together stretched from the north-west of Lough Neagh down the Bann valley nearly to Coleraine. This area was described by Sir John Davys, the Irish attorney-general, in 1607 as 'well nigh as large as the New Forest in Hampshire and stored with the best timber in Ireland'. Of his journey through these oak woods he wrote to the Earl of Salisbury, 'The wild inhabitants wondered as much to see the King's Deputy as the ghosts in Virgil wondered to see Aeneas alive in Hell'[3].

In contrast to these compact oak woods were the woods between south-east Lough Neagh and Strangford Lough. The westerly part of this area was Killultagh, fairly dense oak wood, but eastwards of Killultagh Kilwarlin and McCartan's Country was a countryside of shrubby marshland between the drumlins, with patches of trees on the drumlins' sides. In the Dufferin oak wood again appeared on the boulder clay. The woods here were described in 1602 as having the fairest timber trees in Ireland[4]. Immediately south of Lough Neagh lay Clancan, a lowlying fen with alder and willows, and on the drier islands, oak. Further south Clanawle had denser woods.

Donegal, much of it mountain or covered with poor boulder-strewn soil, does not appear to have had very extensive or thick woods. The relative lack of timber is emphasised by the fact that seventeenth-century planters in Donegal and Tyrone were granted timber from Londonderry. However, the woods south of Mulroy Bay (the Kilmacrennan district) were considered sufficiently important to be reserved for the crown. The woods of oak, hazel, holly, and mountain ash that clothed the slopes of the Derryveagh mountains were the last refuge of the red

Fig 1. When this map was made, Killarney was already famous for its beautiful scenery, woods and staghunts. Not all the woodland, even at that early date, was natural. Ironmasters and other exploiters had taken their toll of the oak, holly and arbutus, and planting had commenced early in the eighteenth century.

deer in Donegal[5] in the late nineteenth century as they were the refuge of woodkernes after the O'Dogherty rising of 1608. A good part of the western shore of Lough Swilly was wooded northwards from Letterkenny with oak, ash, birch, and elder.

Lough Erne was in a very remote part of Ulster at the end of the sixteenth century. As late as 1764 the region was described as a rural Venice where 'the visible and broken parts of the surface appear like so many pieces of water irregularly laid out among the rising woods'. The western shore north of Enniskillen to the mouth of the Arney river was so wild in 1700 that it was 'scarce inhabited by any human creatures but the O's and the Mac's who pillaged all who came their way'. Anything irrecoverably lost was said to be 'got beyond the Arney'[6]. While much of the woodland was oak there was a good deal of ash-oak association on the calcareous soils. As the land rises from the Lough Erne basin the oak-ash woods gave place to an ash-hazel association on the drier ground. Probably also many of the bogs in the area are fen rather than peat. Ash is usually succeeded by beech on chalky soils but beech was not indigenous. Yew, which also thrives on calcareous soil, is found on the islands in Lough Macnean.

In north Sligo the Leguy woods of oak, hazel, yew, and holly lay between the lower slopes of the Ox mountains and the sea, and stretched to Lough Gill, whose twenty-three islands, including the Lake Isle of Innisfree, were wooded. The Loughs Gara, Key, and Arrow were surrounded by woods, in which Sir Conyers Clifford was defeated by Irish forces in 1598. They were described as 'tall, thick woods' but by 1633 they had been despoiled by tenants who sold the timber in Sligo town[7].

The greatest woodland area in south Mayo lay on the shores of Lough Mask and ran northwards to Castlebar. This wood was chiefly hazel and alder mixed with thorn on islands subject to flooding, but the timber on the side of Lough Mask along the foot of the Partry mountains was oak that even in the nineteenth century was only kept in check by cattle grazing. The plain of Mayo was largely forest-free, but with some wood

The Distribution of Woodland, 1600-1800

surviving along its western margin until the mid-seventeenth century.

It is difficult to assess the amount of woodland along the strip of country between Lough Corrib and the western coast. According to tradition one could walk on the tops of the trees from Letterfrack on the west coast to Galway town. Certainly in the mid-seventeenth century there were pockets of woodland between Galway and Oughterard. Along the southern coast of Galway there must have been wood to feed the ironworks of Screeb, Lough Furnace and Doonmore. In the early nineteenth century it was remarked that Connemara possessed extensive shrubby timber (of oak, birch and hazel) on almost every dry knoll or cliff and a little care and protection from cattle would have provided valuable trees[8].

Between the rivers Suck and Clare lies a low range of hills under 500ft which carried at least isolated patches of an oak-ash association, and around Monivea there was a good deal of birch. Between these hills and the woods on the Shannon lay open ground. O'Sullivan Beare before the battle of Aughrim in 1602 lamented, 'Behold the plain lies far and wide before me without any opposing bogs, thick woods or any other place of retreat in which should we fly we could hide ourselves'[9].

North-west Clare was notoriously barren: Cromellian troops complained that there was neither wood to hang a man, water to drown him, nor earth to bury him. Forests began to the east of this region and stretched in a long belt from north of the Clare-Galway border near Loughrea to the Shannon estuary. Around Lough Gara was 'wood of great store of oaks' and 'tall woods and good store of sapling timber'. It was said that a wild cat could have walked on the tops of the trees from Lough Cutra to Creigeen[10]. Just south of the Galway border lay the Wood of Suidan where O'Neill, Earl of Tyrone, sheltered at the end of the sixteenth century when he raided Thomond and Clanricarde.

A narrow belt of wood lay along the Shannon as far north as Killaloe, its remnants being cut in the eighteenth century by Charles Carr, Bishop of Killaloe 'that he might profit by the devastation'[11]. At Killaloe the Shannon widens into Lough Derg, whose shores were almost completely forest-clad, mostly with dense oak wood. By the mid-seventeenth century much of it had been reduced to isolated patches, especially east of Nenagh where it was recorded that woods only fit for fireboot, plowboot, and cabins remained. From Killaloe woodland stretched up the sides of the Arra hills and into the valleys of the Slievekimalta mountains. The north and north-west slopes of the Silvermines were wooded from Cappagh White to Roscrea.

From Athlone to the Shannon's source there was a good deal of woodland, particularly in the land adjoining Loughs Ree and Allen. On the westward side of Lough Ree lay the Fews, which stretched westwards to the 500ft contour. Above this altitude the oak gave place to hazel scrub. The densest of the Shannon's woods lay north of Bofin Lough. Here were 'thick tall woods' that were eventually used up in the Boyle, Arigna, Drumsna, Drumod, and Drumshanbo ironworks. This was the country of MacDermot Roe (late seventeenth century) of the Coill Conchobhair or O'Connor's Wood, which gave MacDermot his title of Lord of the Woods.

An isolated wood in west Roscommon near Ballinlough was so extensive that it took Donnell O'Sullivan Beare and his men a whole night to pass through.

In the south of Limerick, woods lay against the north-facing slopes of the Mullagharerik mountains from Feenagh westwards to Broadford, and continued to Newcastle through Ardagh to Shanagolden and Loghill on the coast, where there were ironworks. The southern section of these woods, known as Clonish, was the gathering place of James Fitzgerald when he rebelled in 1579. In the northern part the Earl of Essex treated with Tyrone in 1599.

Woods stretched eastwards from Rathkeale to link up with the long tract of forest that lay between Charleville and Kilmallock in south Limerick and reached the Shannon at Pallaskenry. This forest occupied the valley of the River Maigue,

The Distribution of Woodland, 1600-1800 45

called by the Elizabethans the May or Maie, and the military reports of the period are full of references to this 'great wood'. Here Sir Peter Carew was killed in 1580 after ignoring advice and entering a wood that was filled with Irish musketry.

Four miles south-east of Kilfennane over a spur of the Ballyhoura hills lay the wooded Daragh-Glenroe, and eastwards of the Ballyhoura hills proper lay the beginning of the wood of Aherlow which continued into Tipperary. In the early sixteenth century Aherlow had been the chief fastness of the Geraldines; by the mid-seventeenth century about 1,000 acres remained of this wood, which had been described in 1580 by the Lord President of Munster as follows: 'It containeth in length three miles, in breadth six miles, distant from Limerick sixteen miles'[12].

Northwards of Croom there were intermittent woods and bogs to Limerick town. From Limerick the woods reached to the foot of the Slieve Felim hills skirting the northern edge of the Golden Vale to Cappamore and Doon.

In the east-west valleys of Cork and Kerry lay mile after mile of forests that were to enrich the Boyles, the Pettys, the Whites and, for a short period, Sir Walter Raleigh; forests which in the first part of the seventeenth century were to cask nearly all the wine that France and (to a lesser extent) Spain would produce, which would float as the hulls of many of the East India Company's ships, which until the mid-eighteenth century would fill the insatiable furnaces of the ironworks that lined the river valleys, and which would provide the bark for the tanneries of Killarney. In the north part of Kerry, woods lay along part of the southern estuary of the Shannon from Leck Point to Ballylongford and were the haunt of prodigious numbers of woodcock. From Listowel there were woods along the foot of the Stack mountains to Tralee, and from Tralee the woods extended eastwards up the Maine valley to beyond Castle Island, known in the sixteenth century as The Island, and westwards along the shores of Tralee Bay and Brandon Bay to Brandon mountains. Woods also lay in the upper reaches of the Feale river in the

Stack mountains. Farther south, about 3 miles north of Tralee were the Sliabh-mis and Glanekinty woods of oak, holly, hazel and birch. The Earl of Desmond was beheaded in 1582 in woods 6 miles beyond Tralee.

On the peninsula between Dingle Bay and Kenmare River Bay many of the valleys that cut transversely into the mountains were forested; these valleys included those of the Carra, the Finnthy, the Inny, and the Blackwater rivers, all of which supported ironworks. At the head of Kenmare River Bay, where Sir William Petty set up his ill-fated English settlement, a number of wooded valleys converged. Further west the valleys of the Sheen, Cloonee, and Glanmore were wooded, the former two containing their own furnaces and the woods of Glanmore supplementing the fuel supply of the furnace across the estuary on the Blackwater.

In the west of county Cork, as in Kerry, the valleys of the Adrigole, Glengarriff, Coomhola, Bantry, and Roaring Water were denuded of their oak, birch, and arbutus to fuel the ironworks lying at the mouths of each of these glens.

In east and central Cork the valleys are wider and longer, and while it is certain that those of the Blackwater, Lee, and Bandon were wooded along most of their courses, it is difficult to assess the width of these tongues of woodland. Those of the Bandon appear to have stretched almost from source to mouth. The woods around Bandon town were to make Bandon tanning famous, but long before this these woods of oak and birch forced an English army to make a detour by Kinsale and Timoleague when marching to the siege of Dunboy in 1602. Part of the woods along the 'Kinsale river' were supposed to be reserved to the crown and in 1611 were earmarked for the navy; but the following year the East India Company founded a settlement of 300 English at Downdaniel, beside Inishshannon, to work at shipbuilding and iron-smelting and paid £7,000 for the surrounding timber[18]. By 1632 the Earl of Cork could write, 'The place where Bandon Bridge is situated is upon a great district of the country and was within the last twenty-four years a mere waste bog and wood serving as a retreat and harbour to woodkernes, rebels, thieves and wolves and yet now (God be praised) as civil a plantation as most in England'.

In the Dunmanway basin, at the head of the Bandon river, considerable wood remained until the eighteenth century. Soon after 1700 £8,000 was offered for the woods there including 'the blocks and stumps that lie in the coppice, because the young growth is very promising and like to come to considerable value!'[14]

In the Lee valley the woods began at the Kerry-Cork border at Lake Gouganebarra and reached almost to Cork, which supports a tradition that a squirrel could have hopped from bough to bough from Killarney to Cork. The Gouganebarra depression was completely denuded of its timber by the early nineteenth century, though a great deal of oak, birch, alder, holly, and other timber survived in the surrounding valleys. An Irish peasant declared of the Gouganebarra woods, when a traveller remarked that it was a pity that the hills round the lake were not planted, 'Planted Sir, why it wanted no man's trees—it was all wood once. A squirrel could have hopped without touching the ground from oak to oak, and from birch to birch, from Inchigeela all along here and up into the pass of Camineagh and so across the hills into Kerry and until you came into Glen Flesk . . . A greedy man here called these trees his own, though the saint, even Saint Fin Barry himself, had surely the best right, he cut them all down'. The despoiler lost his money and his character, 'little better could happen to the chap that would turn to filthy lucre the holy wood of Gouganebarra'[15].

Tributary valleys of the Lee—the Sullane, Shourragh, and Dripsey and part of the Bride valleys—were wooded, the timber of the last named being used to burn lime. Much of the Lee's woods, especially in the lower reaches of the valley, were intermixed with bog, but on the drier parts there were woods of oak, ash, and birch. Those west of Macroom were valued at £4,000 in the mid-seventeenth century, and part of those on the middle reaches of the Lee at its junction with the Dripsey at £850. They were described as suitable for shipbuilding but as of little immediate use because of the difficulties in transporting timber[16].

As on the Bandon estuary there was an attempt on the part of the crown to retain wood for the navy round the Lee estuary. However, by far the greater part of it went to feed Boyle's numerous ironworks or was turned by him into pipestaves.

The Blackwater, the third of the trio of west-east flowing rivers, like the other two, was wooded for the greater part of its course. It was suggested that the woods at Mallow should be reserved to the crown as the river was navigable from there to its mouth but most of them fell into Boyle's hands. The wooded Araglin valley belonged to Boyle's father-in-law, Sir Richard Fenton, who supplied Boyle with fuel for two furnaces at the mouth of the valley. The woods in the Kanturk valley survived longer than those of the Blackwater. In 1655 action was taken to prevent the stripping of oak-bark from living trees for use in Kanturk tanyards[17]. In the next year it was suggested that the best course would be to get thirty or forty oxen, 'and set up some ploughs and cut down Kanturk'. Further east the slopes of the Nagle mountains that face the Blackwater were wooded.

In the north of Cork it was wooded along the border at the Allow river, around Charleville (the Great Wood of Kilmore), and around Kildorrery. The Great Wood of Kilmore, which continued up through Limerick, was one of the strongest barriers against seventeenth-century penetration of the area by English forces.

On the lower Blackwater, and its tributary, the Bride, woods at Condon, Strancally, Lismore, and Lisfinny within 4 miles of their banks were reserved to the crown for shipbuilding but ultimately passed into the hands of the Boyle family. East of Youghal the valley of the Licky was wooded and eastwards to Mine Head.

The woods that began at Dungarvan stretched westwards to meet those of the Blackwater and northwards to the valley of the Colligan. Woodland was also found in the upper reaches of the Mahon around Kilmacthomas.

Woods lying to the south of Waterford town continued up the valley of the Suir to the Tipperary border and reached their greatest width round Portlaw in the Wood of Kilconish, where they were found as far westwards as the Comeragh mountains. Here they probably filled the narrow gap between those mountains and the Monavullagh hills, and continued down the valley of the Nier. At the Gap, properly known as Barnakill, Sir Charles Vavasour and his army met woods of oak, ash, and birch in 1643. The woods on the Suir around Portlaw stood well into the seventeenth century. In Clonegar parish along the southern banks of the Suir 'a natural wilderness of tall, venerable oaks' was still standing in the mid-eighteenth century. Arthur Young remarked on them at the end of that century, 'The whole wood rises boldly from the bottom, tree upon tree to a vaste height, of large oak'.

The Barrow appears to have been wooded from its source in the Slieve Bloom mountains to its estuary, which it shares with the Suir. At Athy two monasteries were built in the fifteenth century in clearings on either side of the Barrow at the entrance to extensive woods. In the mid-eighteenth century woods at Monasterevin, which had long been a retreat for woodkernes, still offered shelter to lawbreakers.

The River Slaney was wooded from Tullow to its mouth. The greatest extent of wood was between Enniscorthy and the Carlow border—the Wood of Coillaughrim, filled with fallow and red deer. The area was described as being 'adorned with goodly woods for main timber fit for building and for pipe staves and barrel staves'. At the time of the Civil Survey, while the acreage of wood throughout the county was considerable, the main body of trees was found in the Slaney valley around Newtownbarry (Bunclody), where there were nearly 10,000 acres of wood and 600 acres of underwood.

The woods of the Slaney extended north-eastwards to Wicklow. On the borders of Wicklow and Wexford were the famous oaks of Shillelagh that had supplied the roofing for Westminster Hall, of which it was said in the 1630s 'no English spider webbeth or breedeth'. It was estimated in 1606 that the Shillelagh woods could furnish the crown with timber for shipping and other uses for the next twenty years[18], and the area was described as 'strong, fast and remote country, the common receptacle and shelter for thieves and ill-disposed members of those parts of Leinster'. This was the area in which the Chamneys had their numerous ironworks but, in spite of their depredations, in 1661 the Earl of Strafford was able to contract for timber from Shillelagh for shipbuilders. However, ten years later when Peter Brousdon was sent from London to scour the country for timber suitable for the English navy, he reported that although Shillelagh was extensive and the timber large and straight enough to make 3- or 4-inch plank, it was so full of shakes and wormholes that he could not advise dealing in it[19]. The Shillelagh area is also of interest in that it was one of two areas that had an organised forestry department during the Commonwealth. The officers were a wood reeve at a salary of £100 a year and four assistants at £25 annually. The other area was in Carlow and Kildare[20].

Clearance of the Shillelagh woods was, for Ireland, relatively slow. The ironworks continued to function almost until the end of the eighteenth century and large parcels of full grown trees and coppice were put on the market until well into the 1770s. By 1830, however, the wood was gone and an old trunk, known locally as the Sprig of Shillelagh, was all that remained of this once famous forest.

Patches of wood lay along the western flanks of the Wicklow hills. In one area, near Timolin, a reward of £100 was given in 1655 to killers of tories, who, after a drum-head court martial, had executed eight of Sir William Petty's English surveyors engaged on the Down Survey[21]. But between these hills and the woods of the upper Barrow valley lay the open Curragh and beyond the Curragh was the Bog of Allen. To the north of the Bog of Allen woods were found in isolated patches westwards

Page 51: *(above)* This edge-wheel, photographed in county Donegal, was used to break flax. Similar wheels were utilised in crushing bark. The axle, a tree trunk of ash or oak, joined a pivot to the stone wheel; *(below)* a twentieth-century cot used for transporting timber and cattle on Lough Erne near Lady Brooke Bridge. Although called a cot it is really a plank boat; the original cots were hollowed from a single oak trunk.

Page 52: (above) Meeting of the Waters, county Wicklow. Castle Howard on the right dominates the glen and on the left fashionably dressed men and women enjoy the rural scenery in this mid-nineteenth-century print. It was of this spot that Thomas Moore wrote in his youth:

'There is not in this wide world a valley so sweet,
As that vale in whose bosom the bright waters meet.'

(below) Killarney Lake from the Kenmare road. A mid-nineteenth-century photograph showing a small pine plantation between the walled road and the lake.

from county Dublin; in the upper reaches of the Boyne Water they fed the Clonard and Sarney ironworks. It was in these woods that Cromwellian soldiers killed 140 Irish in 1650. But the 'Great Wood of the Picts', near Tara, in which Robert Bruce bivouacked, had disappeared.

By the mid-seventeenth century Westmeath was deficient in nothing 'except only timber of bulk with which it was formerly well-wooded'. At the beginning of the century a strip of woodland ran from Lough Sheelin in the north of the county to Kilbeggan on the southern border, where there was an ironworks, with a break between Loughs Owel and Ennell. In the extreme west there were woods about Lough Ree.

In the early nineteenth century there were the remains of several ironworks in the hills in Louth. Their presence indicated that woods were more extensive in Louth than other available evidence suggests.

Possibly Longford did not have very extensive stretches of wood. Only two ironworks are known to have existed, both on the north-west side of Lough Gowna. A description of the county in the late seventeenth century only mentions wood and scrub in the baronies of Clanhugh and Fermoyle (Cala). A survey of 1618 gives the total amount of profitable timber in Longford as 8,400 acres, with 12,500 acres of unprofitable wood and bog. Of this acreage 5,000 acres of profitable and 9,300 acres of unprofitable wood lay just to the north of Longford town[22]. There was also some wood in Killashee parish on the east side of Lough Ree and some in the Inny valley near Ballymahon.

In 1612 during the plantation of Cavan it was claimed that there was no timber in 'the Cavan' for necessary building, and application was made for timber out of the crown woods nearest to that place. In general the woods lay on the periphery of the county. In the extreme west the Shannon valley was wooded where the river enters Lough Allen, and so also was the southern shore of Lough Macnean Upper and the valley of the Swanlinbar. From the border north of Belturbet, woods stretched at least to Ballinagh. At Belturbet in 1611 two English teams

D

of horses were continually employed in transporting felled oak. The northern and eastern shores of Lough Sheelin and round Kingscourt carried timber. South of Cootehill were woods of oak, birch, and alder.

Monaghan was probably better wooded than the available evidence would suggest. At the beginning of the nineteenth century the uplands were described as 'till lately well wooded', and stunted underwood was frequently found at the foot of the hills. Around Lough Corry, for example, there were over 400 acres of scrub, and the Ordnance Survey recorded that within memory the hills were covered with thorn, hazel, birch, and alder of great size and value, but it was cut on the expiration of leases and used for domestic fuel. There were also woods on the northern borders of the county on the eastern side of Slieve Beagh.

Between Slieve Bloom, the Silvermines, Slieve Felim and Galty hills, the land lies for the most part between 300 and 600ft above sea-level; woods lay on the slopes of these hills with isolated patches on the intervening lowland.

By 1700 the most extensive woods lay in the north-west and south-west of Ireland, that is in the remote valleys of peripheral regions. Difficulty of transport in areas without ironworks obviously accounted for the survival of some woodland: for example, in the upper Lee valley the cartage of timber involved impressing tenants to repair roads, for 'without mending those rocky roads it was impossible to have even sold those woods'[23].

Lack of adequate roads in western Kerry necessitated the transport of timber on horseback. On the other hand certain areas could transport timber by river: the Shillelagh wood was brought down the Slaney, Roscommon wood down the Shannon, Leix and Kilkenny wood down the Barrow, Glenconkeyne wood down the Bann, and so on.

In Munster, that is in the south-west of Ireland, some wood survived in the valleys opening into Bantry Bay and Kenmare River Bay and round Killarney. It is significant that the Whites and Pettys, who were big landlords in the region, were able to continue their ironworks until the middle of the eighteenth century. South of the Shannon estuary a remnant of the Great Wood of Clonish survived in the upper valley of the Owveg. Of the woods on the Blackwater, which in 1600 had stretched from mouth to source, all that remained was a patch on the northern slopes of the Nagle mountains. Likewise the formerly extensive woods of the Barrow-Nore valleys were almost completely destroyed, though sufficient timber remained to support the Mountrath ironworks until the second half of the eighteenth century.

The Central Plain was almost bare of timber but on its northern edge beside Lough Allen woods remained until the middle of the eighteenth century, when the local ironworks closed for lack of fuel. Of the Leguy woods on the north side of the Ox mountains in Sligo enough remained to support the Screen ironworks until 1768. On the south side of the mountains during the Penal Laws period mass was said in the shelter of a wood at Largan. The woods at Foxford and Mullinamore on the side of Lough Conn were sufficiently extensive to make it worthwhile for the ironmaster Rutledge to transfer his works there from county Sligo.

In Ulster the main areas of woodland in 1700 were found in the Lough Erne basin, on the fringes of the highlands of Donegal, in the Sperrin valleys, on the northern shore of Lough Neagh, and in the Glens of Antrim. To the south of Lough Neagh it is probable that most of the wood had been cleared: a survey of church lands made in 1703 shows that on such lands, at any rate, only brushwood and stumps were left and that numerous orchards had already been planted and were well established.

In the lower Bann valley and on the north side of Lough Neagh some wood still stood. Tongues of woodland extended into the Sperrins and as late as 1700 the congregation of Draperstown in the Moyola valley was very small because of the wood[24], and the local ironworks continued until about 1760.

On the western flanks of the Sperrins part of the valleys of the Roe, Faughan and Mourne were wooded, and the woods of the upper Roe were not cut until 1770[25]. In Donegal the main pockets lay west of Lough Swilly and in the Finn valley. On the Antrim coast much remained: in Glenariff the wood was sufficiently dense to make the recovery of cattle who strayed into it difficult. But the Lagan valley timber was gone except for a little near Belfast and in Collin Glen.

Over the country most of the larger areas which had been wooded in 1700 were still wooded to a certain extent in 1800, though the woodland may have been reduced to scrubland. The main differences are to be found in the upper Shannon basin and in Kerry. The greatest amount of woodland was probably on the eastern side of the Wicklow hills. Arthur Young described the Dargle Glen (near Bray) as containing 'one of the finest ranges of wood I have seen anywhere' and Wakefield wrote of the vale of Arklow that 'the extent of the woods induced me almost to imagine I was in the midst of one of those immense forests seen only on the continent'.

Such praise for woodland in Ireland at the beginning of the nineteenth century was indeed rare. Only too often there was a lament for the treeless countryside coupled with an exhortation to plant. So it is pleasing to find at least one traveller who could bestow unstinted praise on a wooded glen. The Rev Caesar Otway was able to write of Glengarriff on the southern coast of Kerry, 'In every indenture, hole, crevice and inflexion of those rocks grew a yew or holly; there the yew, with its yellowish tingue; and here the arbutus with its red stem and leaf of brighter green . . . I know not that I ever read of such a place, so wild and beautiful'[26]. So enchanted was he with Glengarriff's beauty that he declared he 'would as soon have gone through Italy and passed by Rome as have missed the glen'.

(See pages 155-6 for notes)

CHAPTER 3

Timber in Industrial Processes

It may fairly be said that the passing of the ownership of the Irish woods into the hands of the new British settlers inaugurated a sort of industrial revolution in the country. During the sixteenth century the most important item exported from Ireland was fish, with salted hides as the second staple export. Exports of lesser importance included wool, woollen cloth, linen, foodstuffs, and, after the middle of the century, small quantities of staves.

Imports into Ireland included large supplies of iron from Spain and from England. The exploitation of the forests in the seventeenth century altered this pattern of imports and exports. The manufacture of staves for export became a major industry, pig and bar iron were made in sufficient quantities for home use and a little was exported, and a trade in tanned hides supplemented the old trade in salted hides.

During the eighteenth century the supplies of local timber were insufficient to meet the demands of coopers and tanners; timber and bark were imported to augment local supplies and the export of timber products ceased. Although the production of Irish iron continued it became insufficient for Irish needs and iron was again imported as it had been during the sixteenth century. Other industries, based on local timber, which developed in the seventeenth century were the construction of house-frames, shipbuilding and glassmaking.

The planting of Ulster by English and Scottish settlers during the first decade of the seventeenth century involved the building of many new towns and villages. Consequently in that part of Ireland there was a demand for timber house-frames. But the rebuilding and extension which took place in Dublin after the Restoration was in stone and brick. Belfast, too, by the mid-seventeenth century was looking not for native oak for the wooden parts of its houses but for imported deals from eastern Europe.

Woods near the sea or a navigable river could be used for shipbuilding. Though it never became a major industry, shipbuilding experienced something of a revival at the beginning of the seventeenth century. Until the 1580s Irish ports had a large number of small ships trading with the continent and with Bristol and Chester. During the last two decades of the sixteenth century a great many Irish boats were sold to Spain and only Waterford and Wexford were left with much shipping[1]. Wexford was forced to use small local boats because the bar at the harbour mouth prevented most English and foreign ships from entering the port. During the following two centuries many small ports round the Irish coast were building ships, usually of under 100 tons and often under 50, but the industry gradually died out in most places.

Glassmaking, which can be carried on with either coal or charcoal for fuel, never developed much outside some coastal towns; there were a few glassworks inland, half a dozen maybe, but they had a short life. This may have been partly because of the high wages which a skilled glassmaker could command.

The way in which timber was used in a district was largely determined by geographical position. Wood near a navigable river could most profitably be made into staves, either for export, or, in south Cork, to cask the animal products of the

Timber in Industrial Processes

provision trade. In remote inland districts, or even inaccessible coastal regions such as the river valleys of west Kerry, the cost of transport made the movement of timber uneconomic and it was converted into charcoal for smelting iron ore.

In this section the industries have been arranged according to the importance of wood in the final product: the casks of the cooper were all wood, ships and houses were partly of wood, the tanner used only bark as part of his tanning process, and in glassworks and ironworks wood was simply fuel.

COOPERING

Of all the industries based on wood, coopering was most completely dependent on timber; every part of the finished article was of wood. The coopers' products—staves and barrels of all sizes—formed a large part of the Irish timber trade during the seventeenth and eighteenth centuries, so an account is given here.

While there are many standard sizes of casks the three that were important in the timber trade were pipes or butts, hogsheads, and barrels. These large containers held 105, 52, and 32 Imperial gallons respectively.

Essentially, the work of a cooper was to produce the two different shapes of wood which are combined in a barrel: the curved staves forming the sides, and the flat pieces fitted together to form the circular top and bottom, technically known as the headings. The staves were bound into shape with hoops, which until the early nineteenth century, when iron hoops came into vogue, were of hazel or willow.

The number of staves in each size of cask was fixed: a pipe had twenty-six, a hogshead twenty-five, and a barrel twenty-three staves. Casks for containing beer or porter, which continues to ferment after it is casked, were made of thicker and consequently heavier staves than those for wine and spirits. The weight of the individual staves in each type of cask is given below.

	Weight of staves in lb	
	Wine and spirits casks	Beer casks
Pipe	8	10.8
Hogshead	4.6	9.8
Barrel	3.1	9.3

There was specialisation within the trade itself into wet coopers who made casks for the provision trade and for brewers and distillers, and dry coopers making barrels for oats, flour, etc. Small bowls and basins were made by turners, as distinct from coopers, who, because of the relative scarcity of elm and beech, relied mainly on ash. Partly because oak was the most easily available wood, staves were originally of oak and they continued to be made of oak for wine and spirits and beer barrels when other woods became available. In the eighteenth century native oak was supplemented by Baltic and American oak. In southern Europe the sweet chestnut was preferred for wine casks, but in Ireland the tree never became more than an occasional exotic and its wood was of no general importance.

The meat, butter, tallow, and fish which formed the basis of the Irish export trade in provisions from the mid-seventeenth century onwards was casked in small barrels. After 1715 only oak or ash staves were allowed in the construction of casks used as containers, sycamore was added to the list of acceptable woods in 1732 and bog timber was specifically mentioned as unsuitable. More latitude was allowed with casks for tallow in 1723 when the use of beech, willow, or birch was legalised[1].

In England coopers were considered to have made heavy inroads on local timber. 'I do not know of consumption of oak timber greater than coopers' ware, particularly in the cider country . . . for which purpose much fine timber, fit for the first rate ships is cut . . . It is cut from the clear part of large oak trees', testified a witness before a British House of Commons Committee in 1792[2]. But in Ireland coopers seem to have escaped the opprobium directed against tanners and ironmasters, despite the large quantity of wood they used.

The Dublin Society, which showed so much concern for many aspects of Irish manufacture, does not appear to have taken any direct interest in coopering, probably because it was a well established industry, but in its endeavours to foster fishing in the north-west and encourage the export of barrelled herrings it offered premiums for the erection of houses for coopers to live in[3]. These premiums were not claimed.

In an age which relied greatly on wooden containers coopering was of course an extremely important activity. There was a Coopers' Guild in Dublin by the time of the Restoration and between then and the end of the century seven foreigners, probably Huguenots, were admitted to the Guild. The Cork Company of Coopers was formed in 1690 and received its Charter in 1701. Five years later the Guild was protesting that in spite of the Penal Laws catholics were exercising the trade of coopering in the city[4].

Cork was the centre of the export trade in provisions and consequently coopering was an important activity there. In 1787 there were forty-six master coopers as against only four timber merchants, and ten years later the number of master coopers had risen to sixty-four.

Dublin, the capital, was more concerned with the liquor and dry provision trade, and in 1752 there were only eight master coopers in the city. This number rose to forty-three in 1768 but fell to twenty-five by 1800. It will be noted that there were fewer master coopers in Dublin than in Cork, possibly because a cask used in the export trade would be non-returnable whereas those used in the home trade could in many cases be used for a number of years.

Newry, like Cork, was engaged in the provision trade, though on a smaller scale, drawing its butter for export from as far away as Cavan, Monaghan and Sligo. In 1803, 27,000 casks were exported and by 1819 there were seventeen coopers in the town.

The cooper, even more so than the tanner, was a necessary figure in the small towns outside Cork and Dublin. While pewter or earthenware or the small wooden bowls of the turner could be used to hold small quantities the coopers' ware was an absolute necessity for large containers. If the north of Ireland can be taken as representative of all Ireland then it would appear that a small village or town needed two or three coopers to serve its needs. For example, at the beginning of the nineteenth century, Ballintoy, Dungannon, and Portadown had three coopers each, Banbridge had two, and there was one cooper in each of the villages of Rostrevor, Markethill, Rathfriland, Waringstown and Lurgan.

Finally, Belfast, in 1791 a town with 18,200 people engaged chiefly in textile manufacturing, could support 115 coopers. Apart from those employed in the textile trades, which absorbed the greatest number of workers, the only type of tradesmen whose numbers exceeded those of coopers were carpenters, of which there were 169.

SHIPBUILDING

A very early and primitive type of boat used in Ireland was the native dugout or cot, which was hollowed from a single tree trunk. These river craft were in use long before the English came to Ireland and continued in use until about the middle of the eighteenth century. The story of the cot has been pieced together by A. C. Lucas[1] and his conclusions are that 'down to the end of the seventeenth century the normal craft of Irish inland waters was the dugout canoe', but that because of 'the wood famine experienced by the ordinary people of the country . . . the dugout canoe became extinct in Ireland at an earlier date than it did in many parts of Europe'.

Lucas, who has gathered together a great deal of literary evidence, shows that the dugout was in use on the Shannon, Bann, Barrow, Nore, Suir, Blackwater, Bride, and Slaney rivers, and on Loughs Oughter, Erne, Derg, Shellin, Foyle, Neagh, and Ennell.

The largest cot so far excavated was found at Lurgan, near

Addergoole in Galway, and is about 50ft long. But generally cots were smaller and carried under a dozen men, often only one or two. An account of cots on Lough Erne written in the last decade of the sixteenth century describes them as 'made from one tree . . . being such as would carry ten men the piece'. Boate in the middle of the seventeenth century said they were 'things like boats but very unshapely, being nothing but square pieces of timber made hollow . . . which kind of ill-favoured boats . . . are very common throughout Ireland, both for to pass rivers in, and to carry goods from one place to another . . . even upon great rivers and loughs'.

We find Michael Cole of Enniskillen petitioning the king for the appointment of 'Constable of the Castle of Enniskillen and captain of all the boats and cots that go upon the Erne' in 1665.

Cots on the Shannon at the end of the seventeenth century could carry over sixty men and were used for shipping horses, cattle and timber down the river. When timber was being moved on them about twenty tons could be taken at a time, the trunks being lashed to poles which were probably laid crosswise over the cot[2].

With the passing of the great trees cots ceased to be made and they were succeeded by the conventional plank and rib boat. What must be one of the last references to these primitive craft is found in an advertisement for the sale of timber from Monasterevan deerpark in 1753. The advertisement says that the 'timber may be carried away by cots and boats'[3].

However, the making of cots, constructed as they were from a single tree, cannot have made very heavy demands on the woods. But after 1600, when the colonisation of Ireland was completed and the country opened up for development, the building of sea-going boats was carried out round the coast and occasionally on rivers and inland lakes. Shipbuilding might have played a greater role than it did in the using up of Irish timber if the English government had been able to carry out its intentions. There was much talk of preserving woods near the Irish

coast for the use of the English navy and some attempts were
made to implement the declared policy of reserving certain
woods for the government's use. Philip Cottingham was sent
from London in 1608 to survey the woods on the south coast to
see if any contained timber suitable for the navy, and in the
following year he paid £71 for the hewing and carriage of wood
in Waterford. Two years later, in 1611, 14,200 trees were earmarked for the government, all within two or three miles of the
water's edge and including 7,500 near Youghal, 3,450 near
Cork, and 3,250 near Kinsale[4]. This, however, appears to have
been all that the government succeeded in appropriating for its
own use. The lord deputy, Lord Chichester, suggested that the
Shillelagh woods could be exported to Milford Haven and used
for shipbuilding there, but nothing seems to have come of this
suggestion. An English act of 1621 prohibiting the felling of any
timber within ten miles of the sea or a navigable river applied to
Ireland[5], but was never anything more than a dead letter.

The difficulty in deciding how much shipbuilding was carried
on in Ireland is that although it is possible to find records of the
tonnage belonging to Ireland or to specific ports these lists give
no indication as to the port of origin (Appendix 2). Although a
ship was listed as Irish that does not say that it was built in
Ireland; it could be a prize ship of foreign origin or it could have
been built in England, as many of them were.

On the whole it is probably true to say that Irish-built ships,
both sea-going and those used on rivers, tended to be small—
often under 100 tons. Even in the eighteenth century, when
English ships were increasing in size, Irish vessels remained
small. The 1,016 Irish ships in existence in 1788 totalled only
60,776 tons.

The small ship was usual in the seventeenth century; it was
considered that a ship of 100 tons drawing 6ft of water was the
maximum size safe for use in the largely unchartered and
uncontrolled estuaries of the British coast, and in 1618 the navy
commissioners advised against the construction of ships drawing
more than 16ft.

Timber in Industrial Processes 65

The dimensions of The Greyhound (1652), a ship of 120 tons, gives some idea of the size of a seventeenth-century ship. Her keel was 60ft long, she was 20ft in the beam, and she carried eighty men. An indication of the value and capacity of one of these small ships can be gathered from the captured Orange Tree. She was originally from Dartmouth and was taken as a prize by the customs at Kinsale in 1666 in unknown circumstances. The ship was valued at £95 and her cargo of pipestaves, hogshead staves, and hoops was worth £313[6].

Structural details of a known Irish ship are lacking but the following description of a famous ship, roughly contemporary and comparable in size, must have applied to many 100 ton ships. Drake's Golden Hind, originally The Pelican, was uncovered at Gravesend in 1911 and a description of her has been preserved by B. Carter[7]. She was about 100 tons and was constructed of ribs 16in square set 16in apart. The outer and inner planking of the hull were 6in and 4in thick respectively, and the kelsons, shaped from a single piece of oak, were 16in thick, 6ft wide, and 8ft long.

On the whole, it took about 1¼ tons of timber for every ton of shipping, and oak of between 80 and 120 years growth was preferred. Elm was used for the parts below the waterline on occasions. Any elm used in seventeenth-century Irish ships would have had to be imported. Masts were of conifers, which, as Ireland lacked this type of tree, came first from Scotland (Coleraine was importing pine masts from Scotland as early as 1611) and from the Baltic. Certain parts, such as the beams, ribs, and kelsons, demanded large pieces of timber and one of the navy's problems was finding sufficient knee and compass timber, which formed best on isolated trees such as those in hedgerows where the branches had room to spread without interference from neighbouring trees.

The life of a wooden ship was relatively short, from fifteen to twenty years[8], but smaller ships lasted longer than larger ones because it was easier to season the smaller pieces of timber. Until 1719 timber was curved by charring one side of a plank and keeping the opposite side wet; after 1719 it was rendered workable by storing in wet sand and heating it; and from 1736 timber was prepared in steam kilns[9].

During the first part of the seventeenth century it is probable that not a great many ships were built in Ireland. Certainly Sir William Petty said that the Irish would never build ships so long as they could hire them from the Dutch, and it was observed in 1660 that if the state 'should have to build shipping in this land' it would be at a loss if the Derry woods were not preserved. The dearth of native ships led to a shortage of seamen. Charles II ordered 1,000 able seamen to be impressed for service in the Dutch war, but the Lord Deputy had to write in return in January 1665 that 'so great are the desolations wrought here by the late rebellion that those ports and the creeks and members of them, do for the most part lie waste and the maritime towns in a great measure depopulated of seamen . . . by shipping belonging some to His Majesty's subjects of England, and some to foreigners and strangers Dutch and others'. There was no chance, the Lord Deputy continued, of getting 1,000 regular seamen, so he proposed to 'take on—for supply of this defect—not only others that now use boats and cotts on rivers, but also other parties that are able bodies of men, who in short time may be made seamen'[10].

Lack of native shipping also had an adverse effect on the Irish economy. The Council of Trade in Ireland in 1668 observed that the cost of freight of goods to and from Ireland amounted to £140,000 annually, which 'for want of shipping of our own in this kingdom it is paid to foreign ships'. They made no reference to lack of native timber hindering shipbuilding but suggested that if a cheap supply of jute for cordage and cables could be arranged then the country would lack nothing in way of raw materials save pitch, tar, and masts[11].

It was not, however, a case of no shipbuilding being carried on but that not enough was undertaken. Small yards are known to have existed at the following seaside towns at various times during the seventeenth century: Sligo, Cork, Youghal,

Timber in Industrial Processes

Wexford, Ross, Dublin, Belfast, Groomsport, Carrickfergus, Whitehouse, Glenarm, Coleraine, and Londonderry. All these places, with the exception of Dublin, were quite near to standing timber. The navy also had temporary yards at Kinsale and Haulbowline Island.

Some ships constructed by the East India Company on the south coast during the first two decades of the century were substantially larger than the usual run of Irish vessels. At the Company's yard at Downdaniel in Cork two ships each of 500 tons were launched in 1613 and in that year the Company enlarged its docks. It also had a yard at Limerick. An Irish-built ship belonging to the Company was at Bantam (Java) in 1611. Later in the century, in 1678, two ships were built at Cork, though not by the East India Company. One was of 60 tons and one under that tonnage. There were also three built at Youghal, of which the largest was 100 tons[12].

In the north of Ireland Coleraine had launched several by 1637; some were of 100 tons and ten more, all under 30 tons, were on the stocks. Charle Monche, the Surveyor General of Customs, alluded to these and wondered that 'shipbuilding had continued there so long without contradiction consuming of ship timber, which His Majesty may have occasion to use'. The cutting of crown woods was forbidden at the Restoration, but Sir John Bennett was given permission in 1668 to take 2,000 tons of timber from the Londonderry woods in return for building three vessels of 30 tons each, to be used for carrying mails across the Irish sea[13].

Shipping lists for Belfast and Carrickfergus in the 1660s designate twenty-seven ships as Irish built. The building yard of all the Irish vessels is not given but the named ones are Belfast, Coleraine, Whitehouse, and Glenarm. There were fifteen Belfast-built ships, the largest one of 50 tons and the total tonnage coming to 245 tons. Three came from Coleraine and were 120, 50, and 40 tons respectively. There was one each from Whitehouse and Glenarm of 24 and 40 tons respectively[14]. The importance of Coleraine as a shipbuilding town is illustrated in these figures; Belfast ships at that time were much smaller as she lacked access to the large oaks of county Londonderry which Coleraine could draw on.

A few years later, in 1676, half of the ships registered at Carrickfergus were Irish-built, including one of 120 tons and smaller tonnages descending to 18 tons.

The parliament of 1689, which was called during the Williamite wars by James II, and known as the Patriot Parliament, made certain proposals to encourage shipbuilding in Ireland. It suggested that anyone building a ship of between 25 and 100 tons within the following ten years should receive a rebate of one-eighth customs and excise duty on the cargo on the ship's first three return voyages to Ireland and ships of 100 tons and over should receive a similar rebate on the fourth voyage also. Masters of ships, seamen, mariners, shipwrights, carpenters, ropemakers and ship block makers who emigrated to Ireland were to be excused from all taxes and quartering of soldiers; and free schools to teach mathematics and navigation were to be established in Dublin, Belfast, Waterford, Cork, Limerick and Galway[15]. But James was a lost cause and subsequent parliaments made no attempt to implement the suggestions of the ephemeral Patriot Parliament.

Shipbuilding continued on a small scale in the eighteenth century and apparently very good though small ships were built. E. Halley in his *Atlas maritimus et commerciales*, published in London in 1728, commented, 'The Irish build very good ships . . . Many of our English merchants chose to build here, for foreign trade especially; their oak is very good and they have good store of it'.

Some shipbuilding was undertaken at Belfast during the eighteenth century; records of four or five ships survive, including the *Royal Charles* of 250 tons, which was launched at the very beginning of the century; but it could not be said that the foundations of the great industry that was to arise in the nineteenth century were laid during the eighteenth. Indeed it would appear that shipbuilding completely ceased, for in 1791 there was not a single shipwright among the population of just over 18,000 people. The basis of the nineteenth-century shipbuilding industry was laid by William Richie, who arrived from Scotland in the year of the census mentioned above, bringing with him ten men and machinery and apparatus. By 1811 Richie had built thirty-two vessels of between 50 and 450 tons and several of them were constructed of Irish oak[16].

On the other hand in Dublin, where shipbuilding died out during the nineteenth century, it had continued all through the eighteenth century, though not on a very large scale.

The Dorset of Ireland, which was described in contemporary newspapers as a large ship, was built in 1731 and launched from Ship Buildings near the Strand, the crush at the ceremony being so great that three people sustained broken legs. Dublin had four shipbuilders in 1741, two on George quay and two on Rogerson's quay, but by 1766 the number had fallen to two and from then until the end of the century there were never more than four at any one time.

The shipbuilding and maintenance trade from the 1750s onwards was in the hands of four families—the Kinchs, the Murphys, the Cardiffs, and the Kehoes. William Murphy was in business in 1751 and a little later the names of Richard, John and Hugh Murphy appear. John Murphy was described in 1764 as 'an eminent shipbuilder' and in the same year Hugh Murphy entered into partnership with the unnamed owners of a dry dock as Hugh Murphy & Company. The Murphy family was still building ships in 1794.

Sobieski Kinch began building ships in the 1760s and Alex Kinch ceased business about 1782. Mathew Cardiff advertised in 1768 that he had acquired and fitted a hulk for ship-repairing. He appears to have worked alone until 1783 and then to have joined forces with Kehoe until 1795, when the partnership ended and both men continued separately in business into the nineteenth century.

Outside Dublin proper Sleater and Raymond Portavine were building vessels at Ringsend in the mid-1760s. A large barge built by Portavine for the Pilot Committee was launched in 1767 in the presence of many hundreds of spectators[17].

Like other employers in the 1770s the Dublin shipbuilders had labour disputes with their men. The employees demanded a daily wage of 3s in 1778 to bring them in line with the wages of English shipyard workers, and they threatened to emigrate if their request was refused[18].

The boatbuilding industry, which had been carried on round the coast in small towns during the seventeenth century, was continued into the eighteenth, though knowledge of it is fragmentary. Waterford in 1746 could boast of 'a large and commodious and well built dry dock with sufficient flood gates and every convenience for the reception of all vessels'. Perhaps it was here that *The Dorothy* and her lighter, both of which the Carews of Waterford had built in the 1760s, originated. The total cost for the two was £139, of which £50 was for timber, £6 for nails, £6 for sails, and £7 10s 0d for rope. The sawyers got £2 2s 0d, the riggers £3 4s 0d and the blockmakers £5 9s 2d[19].

Passage, close to Waterford, had a firm of shipbuilders in 1787, and at the same date there were two shipwrights each in Youghal and Kinsale, while Cove had four.

Cork had three shipbuilders in the second half of the eighteenth century. One of the firms, the Allans, had been in business before 1770 and was still functioning in 1789[20].

The names of two shipbuilders in Limerick are known, Anthony Ellis and John Davis. Ellis was building during the 1740s and advertised that he 'built ships at a reasonable rate'. Davis had a newly built 50 ton sloop for sale in 1771[21].

A Derry man was building in Sligo in 1770. One of his boats, with the keel, stern, sternpost, beams, and hudding ends forward and aft of oak, and the rest of red deal, was 42ft long, just over 16ft in the beam, and the hold gave a head room of nearly 10ft.

Two examples may be given of sea-going vessels being built inland on a river's edge. Two ships, one of 80 tons and a larger one, were constructed at Inistioge, on the Nore, in 1740, and boats of about 40 tons burden were built at the end of the eighteenth century on the River Laune where it leaves Killarney lake. These boats were then taken down the Laune to the sea[22].

The difficulty of determining if ships in use in Irish waters were built in the country has already been pointed out. There are, however, figures available for the end of the eighteenth century—between 1788 and 1800 there were 515 ships built in Ireland with a total tonnage of 23,550 tons, which represents an average tonnage of 46 tons. In other words, even at the very end of the period under review Ireland was still building only small vessels. The insignificance of the Irish trade is emphasised by comparing it with figures for England. In England the naval dockyards between 1730 and 1787 used 37,500 tons of timber annually and the merchant marine consumed three times as much. The tonnage of British-built ships belonging to the East India Company increased from 45,000 tons to 79,900 tons in the five years between 1771 and 1776[23].

During the seventeenth century Irish ships were built of local Irish oak, but during the eighteenth century they must have incorporated some imported wood. While it was necessary to build the hull of oak or elm, softwood was often used for other parts, and this softwood had to be imported as Ireland lacked native conifers. The table below gives the numbers of masts, spars, and oars which were imported into Ireland between 1711 and 1800[24].

TABLE 1

SHIPS' AUXILIARY MATERIALS IMPORTED

Date	Masts	Spars	Oars
1711	4	5,400	—
1751	39	5,400	2,880
1761	5	3,960	2,040
1764–9*	74	6,360	1,330
1770–9*	35	6,240	3,768
1780–9*	52	9,120	6,560
1790–9*	70	10,920	5,760

* Annual average

The total number of masts imported between 1788 and 1800, when 515 ships were registered as built in Ireland, emphasises the small size of the Irish-built ships, especially when it is remembered that some of the masts would go to replace broken masts in damaged ships. The cost of masts varied according to size: in 1717 small masts were from £1 to £2 each, middle-size ones £3 to £5, and large masts were between £9 and £15. By the middle of the century the average price was £2 5s od and this price held until the 1780s.

Likewise the price of spars varied according to size: in the early part of the eighteenth century spar booms were on the average 1½d each and small spars ½d each, but by the middle years they averaged 6½d each. Oars averaged about 1d each and later rose to about 8½d each.

There is one last aspect of Irish shipbuilding to touch on and that is the construction of large boats for use on inland waters. The general lack of roads and the poor state of those which existed during this period made water transport extremely important, especially for moving bulky commodities such as timber. Again and again in the advertisements for the sale of timber during the eighteenth century access to river transport is stressed if there was a navigable river reasonably near the timber stand.

Lough Neagh, the largest sheet of fresh water in the British Isles and still forest-ringed in the early seventeenth century, was an easy waterway from the west of Ulster to the east. Essex found it more convenient to build a fleet of ships to take his soldiers from the east side of the lough to the west at the end of the sixteenth century than march them round the northern or southern sides. Some of the Glenconkeyne and Killetra timber was shipped to Belfast across the lough. Sir George Rawdon, writing in 1670, described how he had gone to Glenavy to advise about the construction of a quay and had seen a dozen boats laden with timber at the water's edge[25].

Three boats built on Lough Neagh in the mid-seventeenth century were of 40, 25, and 18 tons, which compares favourably with the sea-going vessels of the period. Possibly one of these boats was built by Martin Johnston, on whose behalf Sir George

Timber in Industrial Processes

Rawdon wrote to Viscount Conway in 1668. Johnston, he explained, wished to buy at the current rate some timber which had fallen in Tunny Park to build a vessel of between 30 and 40 tons. Johnston also had a project to build a pleasure boat but he wanted to pay for the timber after the boat was completed[25].

A Lough Neagh boat of 50 tons which aroused general interest at the time because of the origin of its timbers was built near Glenavy in 1780 from part of the wood of the Royal Oak of Portmore and was given the name of the tree—*The Royal Oak*.

HOUSEBUILDING

One immediate result of the Tudor conquest of Ireland was that many small towns and villages were built to accommodate the new settlers, particularly in the north. The houses in these new settlements were built according to current English fashion, that is they were of the famous half-timbered type. Such houses were not completely new to Ireland, as they had been introduced at the beginning of the sixteenth century, but they became much commoner as a result of the influx of English immigrants at the end of the century. None of this type of house has survived to the present day in Ireland, but they could still be seen at the beginning of the nineteenth century. For example, in Antrim a few were in use in 1837 and in Kilkenny they survived until even later in the century.

A very fine example of a half-timbered house stood in Drogheda until 1824. This house was built in 1570 of oak from Mellifont Park by Nicholas Bathe. The second storey, which was of square oak framing with spandrel pieces, each piece forming a quadrant or segment of a circle, the interstices filled with plaster, projected over the two lower floors. The rooms on the first floor were panelled with carved wainscot. It was considered a particularly fine example of its type and Taaffe the traveller remarked, 'I have seen wooden houses in Pilnitz, Reichenau and other towns of Bohemia and Germany, but none of such curious

Fig 2. This fine specimen of a house built of Mellifont oaks, at the junction of Laurences Street and Shop Street, Drogheda, was erected in 1570 and survived for 254 years.

and elegant, as well as durable workmanship'. It was demolished by Drogheda Corporation, having been for many years suspected of harbouring rats, reprobates and typhus fever[1].

In Dublin the early wattle and clay houses were replaced by half-timbered types in Elizabeth's reign, and these in turn gradually gave place to slated stone houses. There were, however, quite a number of half-timbered or cage-work houses in use as late as 1770—nearly all in Patrick street, Castle street, High street and Wood Quay. The last of them, at the corner of Castle street and Werburgh street was taken down in 1813. Brick-built houses probably first appeared in Dublin in Queen Anne's reign.

To speed the building of Coleraine at the time of the Plantation a small number of wooden frames were imported from England, but the greater part of the wood used in the new towns was local and in Glenconkeyne and Killetra the making of house-frames became a small industry which survived around Bellaghy until well into the eighteenth century[2].

The availability of wood for houses was one of the inducements offered to the City of London to participate in the Ulster Plantation. Each planter in 1610 was allotted 200 good oaks to make timber for such buildings as he wished to erect and in the Charter given to Londonderry City in 1613 it was laid down that the timber in Glenconkeyne and Killetra was to be used only for house construction or other necessary purposes, a condition repeated in the Charter of 1674[3].

According to A. G. Henry 150,000 oaks at 10s each, 100,000 ash at 5s each, and 10,000 elm at 6s 3d each were used in the building of Londonderry (the elm must have been imported). The initial blueprint for the City envisaged 200 houses to start with and a further 300 within six years. At Coleraine 100 were to be put up immediately and 200 more inside the next five years. However, at the City of London's trial in 1633 it was alleged that by 1628 only 114 houses had been put up in Londonderry and that they were only 36 by 20ft, which was smaller than the original stipulated measurements. The correct

Fig 3. Of unknown age when dismantled in 1813, this was the last surviving example of a type of house common in Tudor Dublin. It stood at the corner of Castle Street and Werburgh Street.

number had been erected in Coleraine but they were 'the poorest and weakest that ever had the names of houses'[4].

The Earl of Abercorn had built large wooden houses in Strabane by 1611. These houses were 116ft long and 87ft wide with groundsels of oak, the other parts being of alder and birch[5].

In other parts of Ulster houses incorporating native oak were going up. Lord Chichester cut 600 oaks to build his houses at Carrickfergus and Belfast, and Sir Henry O'Neale took a similar number from the Londonderry woods to build his house at Killylough. The Waring family house, Waringstown, built in 1671, retained its oak roof until 1834, and the local church, completed in 1618, had a massive roof of oak taken from the Clanconnell district[6]. But if the large estate houses were incorporating native oak not all tenants' houses were built so solidly. Montgomery's own house at Newtownards was roofed with oak from his woods, but the cottages on his estate were made of sods and thirty-year-old saplings of ash, alder, and birch, roofed with rushes and wattled with bushes.

A call on Irish oak was made after the Great Fire of London in 1666. At the rebuilding of the city only oak was permitted for roofs, doors, window frames, and cellar floors, and to help meet the demand a Captain Henry Nicall bought a ship, *The Wild Boar*, for £1,000, to help ship oak from Ireland[7].

Although oak was shipped from Ireland to help rebuild London, in Belfast and Dublin softwood was being increasingly used for housing, though in Dublin the use of any sort of wood was restricted in 1670. In that year the Lord Lieutenant decreed that all new Dublin houses were to be built of stone or brick and to be slated or tiled. All thatched roofs were to be replaced within a year and overhanging windows, typical of the old half-timbered houses, were prohibited, though balconies were permitted[8].

At the same time, 1670, softwood was popular for houses in Belfast. 'They cover with deal', wrote Sir George Rawdon, 'pitched and so ordered that they are very staunch and light and the up-top rooms as good as the middle storey'[9].

The first indication that timber was not so universally available for building as it had been came during the Commonwealth, when settlers were permitted to cut trees for house construction only under licence and against security[10]. In the mid-1660s Rawdon declared that as good timber was very scarce in Dublin and was fetching £2 5s od a ton he was going to cut some of his oaks in Tunny Park and send them to Dublin via Belfast. The scarcity in Dublin, however, seems to have been primarily of softwood, whose import was hampered by the Dutch wars. By this time the Dublin builders, like those in Belfast, were putting mostly deal into their houses.

The decreasing amount of timber in Ireland was making itself felt even in the more remote country districts before the end of the century: those areas with adequate wood reserves could congratulate themselves, as a letter written from Portadown in 1682 shows. 'The great plenty of wood which this barony affords', says the writer, 'makes our houses much better than those of other parts where that assistance is wanting.'[11]

To meet building needs timber was imported from the western highlands of Scotland and in an effort to conserve native resources the use of saplings in the wattling of walls was forbidden in 1705. From Ballyclare, county Antrim, came the lament in 1718 that timber 'being so extraordinary dear here the profit will not do much more than defray the expense of building'. This sentiment was echoed from Carrickfergus in 1735. 'But timber is so dear and tenants in the melancholy place so uncertain that it is better to sell the ground than to build.'

It was said in the 1730s that imported timber was so dear that people lived in cabins and lost their cattle for want of wood to make sheds for them to shelter in. At the beginning of the nineteenth century the cabins in Armagh, Londonderry, and Tyrone were being roofed with bog oak and in Leix bog yew was used for the purpose. This is a telling comment on the scarcity of wood for building in some districts, for in the mid-seventeenth century scrappy wood was often slightingly referred to as 'fit only for fireboot, plowboot and cabins'[12].

Most large eighteenth-century houses had wainscotting in some of the rooms, partly as decoration and partly as a protection against damp. Galgorm Castle in county Antrim, built during the seventeenth century, was wainscotted with oak from the Largy and Grange woods, and John Wesley was enthusiastic about the wainscot in the Irish Parliament House when he visited Dublin in 1787. 'The House of Commons', he wrote, 'is a noble room indeed. It is an octagon, wainscotted round with Irish oak, which shames all mahogany.'[13]

Before the end of the seventeenth century wainscot was being imported into Ireland, though on a very small scale. In 1682, for example, 160 pieces were sent from eastern Europe to Ireland. Its import during the last third of the eighteenth century was rather less than might well have been expected in view of the expansion of Dublin during that period. The value and amount of wainscot imported between 1764 and 1793 is shown below.

TABLE 2

IMPORT OF WAINSCOT, 1764–93

	Value (£)	Pieces (in hundreds)
1764–9	1,740	87
1770–9	2,600	130
1780–9	2,660	132
1790–3	300	15

TANNING

Tanning was another industry dependent in the seventeenth century on the local forest resources, as indeed it had always been. Unlike glassmaking or iron smelting on a large scale, tanning was a very old Irish activity. While during the seventeenth century tanning was dependent on the native woods, by the eighteenth century oak bark was being imported, chiefly from England, but also from the continent.

Tanners preferred to strip bark from the living tree and a reasonably competent man could strip a tree in two hours.

If felled oaks had to be barked it was preferable, from the tanner's viewpoint, that it should have been felled in the spring as then the bark was most easily removed.

The amount of bark obtained from a tree varied according to its age: twenty-four oaks of twenty-five years' growth yielded a 12 stone barrel of bark, a forty-year-old oak yielded 9–12lb of bark per cubic ft of timber, and one over eighty years old gave 10–16lb of bark a cubic ft. Abnormally large trees could produce a great weight of bark—the famous Royal Oak in Conway Park, near Antrim, which girthed 42ft yielded £40 of bark in 1780 when bark fetched £6 a ton[1].

Richard Boyle estimated in the early seventeenth century that 8,000 tons of timber would yield him £400 worth of bark, and in 1665 a barrel of bark cost 6s 8d, but unfortunately the capacity of such a barrel is not known. By the beginning of the nineteenth century a barrel seems to have been standardised at 12 stones.

During the eighteenth century imported bark was valued at 6s a barrel in 1720, 6s 6d in 1744 and 7s in 1783. Local Killarney bark cost £2 10s 0d a ton in 1736 and it was the same price at Portumna in 1745. Imported bark was £3 a ton in 1760. The cost of stripping and sawing a ton of bark near Carrick-on-Shannon in 1772 was £1, and the cost of felling, stripping, transportation, 'saving, watching and weighing', a ton of oak bark on the Oliver estate near Ballyorgan in Limerick came to £1 10s 0d in 1777. This bark was then sold at a clear profit of between £6 and £7 a ton. Birch bark was cheaper and was sold at £4 a ton[2].

In common with timber the price of bark rose during the Anglo-French wars and at the end of the eighteenth century had risen to £11 or £12 a ton. After 1800 there was a further increase to £20 and over. During the 1790s, tanners had to accept bark with the outer layer unstripped so that a stone was lost out of every hundredweight.

The tanning industry received a great impetus during the mid-1660s when the export of live cattle from Ireland was prohibited and the number of tanned hides rose from 106,300 in 1665 to 217,000 in 1669[3]. The export of tanned hides continued into the eighteenth century, helped by the growth of the provision trade, but in the middle of the century it declined and the export of untanned hides increased. This was said to be due to the dearness of bark and landowners were recommended to coppice any of their woods valued at under 3s an acre to supply bark for the trade[4]. There was also an export trade in shoes, which for custom purposes were assessed by the pound weight. Their main market was North America, which took 12,700lb of shoes in 1783, and there was also a small quantity sent to Portugal—1,463lb during the same year. The cost in Dublin at this time was 1s per lb for uppers and 9d per lb for soles.

There does not appear to have been any appreciable export of bark from Ireland during the seventeenth century: for example, in 1665, when the export of timber was in full swing, no bark was sent abroad and only 142 barrels to England. The import of bark into Ireland on a significant scale roughly coincided with the beginnings of the import timber trade, around 1710. Between 1719 and 1727 the annual average value of imported bark was £16,100, of which £12,500 worth came from England. In 1736 the value was £18,400, in 1744 it was £15,400, and by 1783 it had risen to £36,000 of which £31,000 was of English origin. The foreign bark came from north Germany and was about two-thirds the price of English bark but it took longer to tan with it. In an effort to reduce the amount spent on imported bark the Dublin Society offered premiums from the 1730s onwards to anyone who could discover a suitable substitute for oak bark, though in 1727 a William Maple had been given £200 by the Irish government for an alleged method of tanning hides without bark using 'a vegetable of the growth of this kingdom'.

The Dublin Society offered a premium of £10 in 1750 to the person who tanned most hides with tormentil roots and £8 to whoever gathered and sold the most tormentil roots to tanners. This plan proved unsuccessful so the Society concentrated on the cheaper German bark and in 1775 offered premiums of 10s

a ton for oak bark imported from Bremen and Hamburg.
A Stephen Roche was awarded this premium on 179½ tons of
it and an Alexander Jeffray on 45½ tons. This particular action
of the Dublin Society predates the proposal of Sir John Parnel,
chancellor of the exchequer, that a bounty of 3s a barrel should
be paid in 1794 on all imported bark.

In face of the general shortage small quantities of bark from
trees other than oak were used—willow, birch, poplar, and ash
among others—and these types of bark were slightly cheaper
than oak.

Because of the havoc tanners could wreak in a wood they met
with hostility from other tree-users. Peter Brousdon, searching
for timber for the British navy in 1670, said that Sir William
Petty's woods in Kenmare were in better shape than most other
woods because of the absence of brogue-makers in the district[5].
An attempt was made to control tanners before the question of
timber conservation in Ireland was mooted. The setting up of a
tannery without a licence was prohibited in 1569 and the
tanning of leather in places unauthorised by the Lord Deputy
was forbidden. Orders were issued by Sir Thomas Phillips in
Ulster in 1623 that tanners were only to buy hides in the open
market, and a brogue-maker could only live in a town or village
'under some English gentleman'. He also tried to enforce the
1569 act. The barking of trees for tanning was forbidden in
1628, and in 1634, when it was declared that the barking of
living trees 'by lewd and mean persons' was more prevalent
than it had been. The 1634 act was renewed in 1705[6].

By 1649 the number of tanners in each county had been
limited and unauthorised people convicted of barking trees were
made liable in 1711 to a fine of 10s, which was to be used
for the benefit of the poor of the parish. Justices of the peace
were authorised to search suspected houses. It will be noted that
the date of this legislation coincides with the substantial
increase in the import of bark. After 1767 it became illegal for
anyone other than a recognised tanner to obtain bark.

Despite the impression given by Sir Thomas Phillips'

restrictions, that tanners were lurking in every wood, they appear to have been scarce in Ulster in the early days of the Plantation. One of the earliest recorded tanners was Walter Hillmane, a burgess of Carrickfergus, who operated a tanyard in 1611 in Gallanagh, county Down. By 1613 there were tanneries at Castle Rowe, Coleraine, and Limavady. Permission was given to Sir Randall McDonnell in 1629 to erect one or more tanneries in the Glens of Antrim. Tanneries were established in Belfast before 1638 and Thomas Waring, mayor of Belfast five times between 1652 and 1666, had a large tannery between North street and Waring street. Previously Thomas's father, John, had set up a tannery at Toome on the north shore of Lough Neagh, but Thomas moved the business to Belfast sometime between 1640 and 1650. Other members of the Waring family, Joseph, John, and Thomas, had tanneries at Derriaghy, near Belfast, and at Lurgan. John Waring obtained the right to bark all woods and underwoods in seven townlands on the Brownlow estate near Lurgan, including the townlands of Derryadd, Derrytrasna, and Derrytagh North in county Armagh, and Ballinderry in south-west Antrim, for twenty-one years at an annual rent of £60. About the same time Philip Shelton's father kept a tanyard at Derriaghy, and in 1669 it was said that tanned leather was plentiful in the Lagan valley, 'the plenty of woods and cheapness of hides' facilitating the trade[7].

Sir William Petty in his *Political anatomy of Ireland* (1672) states that there were 20,000 shoemakers and 10,000 tanners and curriers in Ireland; but his figures may only be reliable enough to indicate that tanning was widespread. Also, the term tanner could be an elastic one: in Cork in the mid-seventeenth century where there were three tanyards the Society of Tanners included chandlers and soap-boilers.

Other places known to have had tanneries at this period were Limerick, Kanturk in county Cork, Bray and Dublin.

During the first half of the eighteenth century tanyards were in operation at the following towns and villages: Ballynastra, Ballyclough, Ballynacarrig, Ferns, Ross, Wicklow, and Dunlavan in counties Wicklow and Wexford; Rathfarnham and Dublin in county Dublin; Edenderry in county Westmeath; Monasterevan in county Kildare; Ballynakill in county Leix; Ballintogher in county Sligo; Thurles and Dundrum in county Tipperary; Killarney in county Kerry; Gort and Creggs in county Galway; Birr in county Offaly; and in the cities of Cork, Limerick, Belfast, and Waterford.

Between 1750 and 1770 tanneries are known to have existed in Ballinaclash in county Wicklow; Cootehill and Killashandra in county Cavan; Ballycastle in county Antrim; Bandon and Dunmanway in county Cork; Kells in county Meath; Ballyboy in county Offaly; Callan in county Kilkenny; Shanagolden in county Limerick; and in Cork, Waterford, Limerick, Dublin, Belfast, and Enniskillen[8].

The number of tanners in Dublin rose from thirty-six in 1768 to forty-five in 1800. They were concentrated in five areas—James street, Cork street, Mill street, Dolphin's Barn and Kilmainham.

A few of the tanners combined tanning with allied trades: there were tanners who were also curriers, skinners, leather merchants, butchers, or glue-boilers; and there were a few who practised unrelated trades such as weaving, bed-making, distilling and brewing. There was also a tanner who combined tanning with cheese-making. The combination of tannery and brewery was not confined to Dublin; the tanyard at Monasterevan had a malthouse attached to it capable of malting 2,000 barrels a year, and the tanneries at Pennywell in Limerick and at Birr were allied to breweries.

There is not enough evidence to classify tanneries by size, though two isolated examples may be given: the Birr tannery could deal with 6–700 hides at a time, and the rent of one in Mill street in Dublin was £50 per annum in 1750, plus £45 for an attached brewery.

Wakefield made two observations on Irish tanyards at the beginning of the nineteenth century. He said that the quality of Irish tanned leather was poor, and that Irish tanners lacked

Page 85: (above) Richie's Dock, Belfast 1805, showing small wooden ships in course of construction fourteen years after Richie came to Belfast from Scotland. His dock was near the present site of the Belfast Harbour Office; *(below)* the Dargle, county Wicklow, a wild valley, thickly wooded with natural oak about 1800. Its beauty was spoilt by the cutting of a road in 1821 to permit George IV's carriage to drive through it.

Page 86: *(above)* The timber bridge at Cappoquin, county Waterford. When this drawing was made in 1834 the bridge was in a dangerous condition, shaking from end to end as a person crossed over it. At that time it was at least two centuries old as repairs had been carried out on it at the Restoration; *(below)* Hayesbridge on the river Avonmore at Avondale. Samuel Hayes, the author of *A practical treatise on planting,* and one of the founders of the Dublin Botanic Garden at Glasnevin, designed and built this wooden bridge on his estate in Wicklow about 1790. It spanned 60ft and rested on stone pillars.

the mechanical means used in England to separate the layers of bark. The poor quality of the leather he attributed to the method of levying duty in Ireland. In England the duty was on the skin itself, whereas in Ireland the duty was levelled on the pit, which encouraged the tanner to put as many skins as possible through the same liquid, and thus says Wakefield, 'the skins are imperfectly prepared, the process being but half performed'[9].

With regard to Wakefield's other statement, that Irish tanners had no mechanical means of separating the layers of bark, they at least had mechanical aids for grinding bark. For example, there was 'an old bark mill house' just north of Wicklow in 1747[10] and tanners' advertisements list bark mills among their equipment. They were probably not very efficient, for the Dublin Society gave a premium in 1794 to Paul Panton, a tanner from Chester, 'for a new construction' for grinding bark, and in 1802 a premium of five guineas to James Lambic for a bark-grinding machine which had earned 'the entire approbation of the principal tanners' in Dublin.

Finally, two advertisements may be quoted to show the equipment of an eighteenth-century tanyard. The first is from Kells in 1763 and the second from Waterford in 1765. The Kells advertisement runs: 'To be let. A tanyard with twenty one pits, three limepits and a quantity of strong ooze with a dwelling house, bark house and bark mill and utensils belonging to a tanyard'. The Waterford advertisement states: 'To be let by William Morris, esq., Large tanyard in Waterford city near the Graving Bank. Fine bark kiln, two bark mills, two leather cellars with a large bark loft over them and a closet. Another large bark house, a beam house and lofts over it, two large drying houses over the scowerings and handlers'.

GLASSMAKING

Two industries in the seventeenth century drew on the forests for their fuel supply—glassmaking and iron smelting. The former was not dependent on charcoal as a fuel, as the latter was for chemical reasons; any fuel could be used in a glasshouse furnace and imported coal from England, or local coal if it were available, was used in glassworks in coastal towns.

The amount of timber used in a glassworks was much less than that fed into an ironworks. At his glasshouse in Ballynageral in county Waterford Richard Boyle used twenty-five cords of small cleft wood a week to make eighteen cases of glassware. This glass he sold for £1 the case, and his glassmaker, Davy, surely one of the highest paid craftsmen in Ireland at the time, earned £2 14s 0d a week making these eighteen cases. Although there were few glassworks and the consumption of fuel was not large it was thought worth while to include glassworks with ironworks in the prohibition of felling for fuel in 1641. However, while the production of glass was not tied to wood as a fuel, it was an asset to have access to ash trees, as the alkali used in processing glass was obtained from them.

The earliest effort to establish a glassworks in Ireland appears to have been in 1588, when a George Longe suggested to Lord Burleigh the transfer to Ireland of thirteen out of the fifteen glassworks then existing in England. One of his reasons was that 'the woods in England will be thereby preserved and the superfluous woods in Ireland wasted, than which in time of rebellion her Majesty hath no greater enemy there'. At this time Longe was making glass at Curryglass in county Cork and taking his fuel from the Drumfin woods that lay between Dungarvan and Tallow.

In 1597 Longe requested a licence for the monopoly of glass-making in Ireland. He repeated the reasons he had formerly advanced to Burleigh and continued, 'Much trade and civility will increase in that rude country by inhabiting those great woods . . . neither in Ireland shall any timber be wasted, there being such mighty places and underwoods that impossible it is to spoil them, continually growing again . . . For example I have kept (a glasshouse) on the end of Drumfin woods . . . There is no sign of waste only the ways more passable'[1].

More glasshouses were erected in the early years of the seventeenth century: Adam Whitty of Arklow in 1609 applied for permission to manufacture glass for ten years; Richard Boyle had a works at Ballynageral by 1618; and a William Tipton charged £8 for the timber frame, doors, boarding, ladders, and stairs of a Dublin glasshouse put up in 1621. The lease of land and wood at Clonoghill and Clonbrone, near Birr Castle, for a glassworks was made in 1623, and this works was still functioning in the middle of the century supplying Dublin with window panes and drinking glasses[2]. An old patent for glassmaking was surrendered by Sir Percival Hart in 1634 and a monopoly granted to him for making black Venetian glass cups and vessels for twenty years, but it is not known which glassworks he controlled. Among the list of Irish grievances presented to Charles I in 1641 was the granting of monopolies for glassmaking[3].

The duty on imported glass in 1665 came to £984 and possibly with the idea of tapping the Irish market for glass Viscount Conway wrote in that year to his brother-in-law, Sir George Rawdon, with proposals to set up a glassworks either in the Lagan valley, probably near Lisburn, or somewhere on the east shore of Lough Neagh. Rawdon, who was stocking his estate on the lough shore with deer and whose main interest was digging a canal from the lough to stock with duck, was against the proposal. They would not, he asserted, see their money back on the venture and the local sand might not be suitable. Conway continued to press the idea and suggested Glenavy as the site; Rawdon then flatly and mendaciously asserted, 'There is no wood near and the carriage of it is very costly'. The proposal seems then to have been dropped[4].

Ananias Henzey, who had 'practised glassmaking in another place these twenty years past' (probably Stourbridge) was in 1670 finding it unprofitable to make glass at Portarlington and appealed to Lord Arlington for help. The fate of this glassworks is not known[5].

After the 1670s no more is heard of inland glassworks using

timber for fuel, or indeed of any inland works, but they appear in the coastal towns using imported coal. The earliest of these, in Dublin, was probably built about 1675 and during the eighteenth century several more were started in the city. The Waterford glass industry began during the 1720s. In Belfast the first glassworks was set up by Benjamin Edwards in 1775, but three years earlier he had run a glass factory at Drumrea near Dungannon, using coal from the local Tyrone coalfield. Once his Belfast glassworks was established Edwards built an iron foundry beside it, and at the turn of the century he built a foundry at Newry. Glassmaking was started at Cork in 1782, and at Ballycastle on the north Antrim coast a glassworks was in production for about twenty years from 1754 onwards, using coal from the local Ballycastle colliery.

IRONMAKING

Iron ore is widespread in Ireland, in veins, bedded and as a bog deposit. Limestone, which is used as a flux in iron-making is also widespread. These facts, allied to the difficulty of profitably selling timber any distance from water transport, accounted for the mushroom growth of ironworks—over 160 of them[1]—all over Ireland during the seventeenth and eighteenth centuries (see Appendix 3).

It was not until the end of the eighteenth century that the technical problems involved in smelting iron ore with coal were even partially overcome, though Dud Dudley, who owned works in the Forest of Dean and was also connected with the Enniscorthy works in county Wexford, was experimenting with coal as a fuel at the time of the Restoration. The technique of coal-smelting was not satisfactorily solved until the nineteenth century, when a hot-air blast in the furnaces replaced the traditional cold-air blast. The various attempts made in Ireland in the eighteenth century to use local Kilkenny and Roscommon coal, and peat, as fuels were unsuccessful; and so, during the period under review, charcoal was used for smelting both in

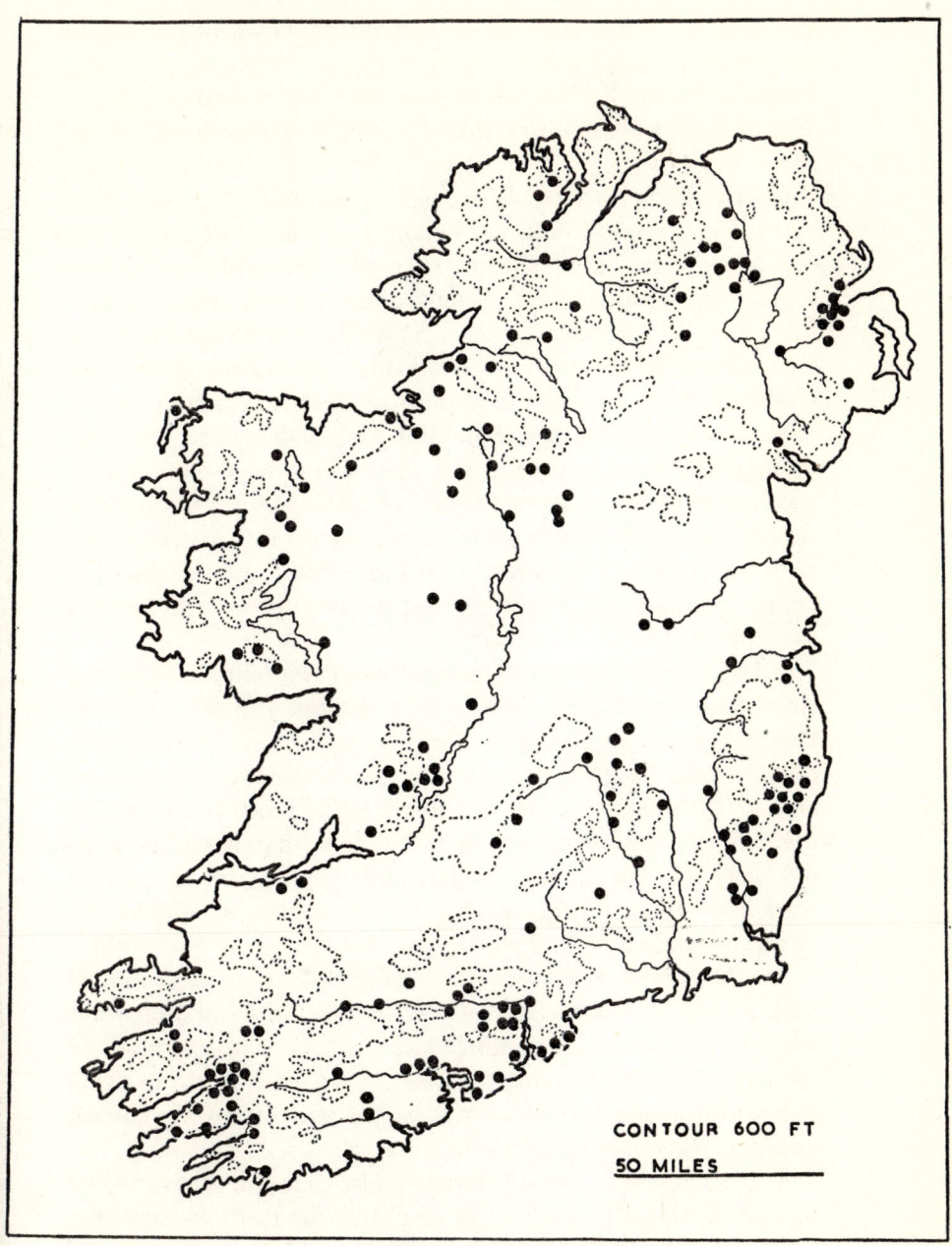

Map 5. Furnaces and forges, 1600–1800

furnaces where the raw iron ore was smelted to make pig iron and in forges where the pig iron was converted to bar iron.

Ideally the best charcoal for smelting comes from twenty-five-year-old coppice oak, and in England the iron masters practised coppicing to ensure a continuous supply. Generally, an acre of coppice gave enough fuel to make a ton of iron every twenty-five years. But in Ireland, except in Wicklow, no such provision for a continuous fuel supply was made, and the life of an ironworks was limited by the supply of readily available wood. Some ironmasters, such as Rainey in Londonderry and Rutledge in Sligo and Roscommon, moved their works from one place to another as the local fuel became exhausted. Other ironmasters closed their works for part of the year; for instance, the Mountrath and Enniscorthy works, both owned by the Poundens at the end of the eighteenth century, were in production for only three months every year.

All the known ironworks except that at Belmullet in Mayo lay close to the fuel supply, because charcoal, being friable, crumbles very easily into useless dust. The ore was brought on horseback to the works, sometimes over considerable distances. As a good deal of Irish ore is of low quality English, Scottish and Welsh ore was imported to mix with it. Cinders, that is residue from old working, which contained considerable quantities of iron were imported from the Forest of Dean, for use in some of the Wicklow and Wexford works.

Gerald Boate, writing in the seventeenth century, recorded that it was almost incredible how much timber an ironworks used. A picture of the amount of charcoal kept in tips near the works is given in an account of the Drumshanbo works in 1770, which states that the works was ringed with heaps of charcoal as big as three Dublin houses[2].

Timber for the works was measured by the cord, that is a loose bundle of branches about 12ft long and about 4ft in diameter. About twenty of these bundles were needed to make a ton of bar iron from ore, or, to put it another way, it took about 2¼

Timber in Industrial Processes

tons of charcoal to make a ton of bar iron. Irish wood was cheap by English standards. In the mid-seventeenth century English ironmasters paid 6s to 7s a cord, whereas in Ireland the price was in pennies rather than shillings; it was not until the end of the eighteenth century that an Irish ironmaster had to pay 7s a cord[3].

Many of the ironworks supported quite considerable colonies of people, and may be compared to present-day industrial developments in rural areas. It is almost certain that most of the workers were English or European immigrants especially brought over as a labour force. Sir Charles Coote had 2,500 workers, English and Dutch, at his various works in Cavan, Leitrim, and Roscommon; Sir William Petty founded a colony of 800 English at his ironworks at Kenmare, on which he expended about £10,000; English families were brought to Enniscorthy after the Restoration; and there were Walloons from Liége at Tomgrany in Clare. Special permission had to be obtained to employ 500 Irish workers at the Mountrath ironworks in 1654 until English workers could be obtained, and the Irish had to live within a musket shot of the works. However, it is clear that Irish workers were sometimes employed: for example, Colonel Brown of Knappagh in Mayo asked for permission to continue to employ Irishmen during the Williamite wars when gatherings of more than ten were prohibited[4].

Comparison of English and Irish wages during the seventeenth century shows that the skilled workers—furnacemen, hammermen, etc—earned about the same in both countries, but the fellers, the charcoal burners, and the carters earned considerably less; and it is possible that these latter, the hewers of wood and drawers of water, were Irish. It is likely that women were employed in such tasks as carrying baskets of charcoal from the tips to the furnace.

The cost of establishing one of these early industrial works was considerable. If land and timber had to be bought £3,000 was needed, but if one had land and timber then £1,000 was sufficient. The land necessary for an ironworks with workers' houses and plots covered between 200 and 400 acres.

The list of Irish ironworks in production at various times between 1600 and 1800 given in Appendix 3 is not exhaustive. There were also works in Donegal, Louth, Westmeath, and Wicklow that have not been traced. Of those which have been satisfactorily identified not many have been satisfactorily dated: the opening and closing dates are known for only a handful of them. However, on the basis of such dates as are available, and excluding Dublin, about sixty belong to the seventeenth century, about forty to the eighteenth, about thirty spanned both centuries, and the remainder are undated. Thirteen works, listed below, are known to have lasted for over a century, but whether they were in continuous production is another matter.

TABLE 3

IRONWORKS IN PRODUCTION FOR OVER A CENTURY

Site	Dates	Years
Enniscorthy	1560–1792	232
Mountrath	pre 1641–1792	151
Clonelly	1611–1758	147
Arigna	pre 1641–1788	147
Araglyn	1625–1770	145
Castlecomer	1635–1770	135
Cappoquin	1615–1750	135
Creevelea	pre 1641–1768	127
Drumshanbo	1640–1765	125
Shillelagh	1641–1767	126
Mount Mellick	1630–1756	126
Boyle	1659–1763	104
Drumod	1695–1798	103

Not unexpectedly the earliest works were near the coast: in Antrim the wooded Lagan valley supplied fuel for the Belfast works, and in Waterford Sir Walter Raleigh had works which passed, together with his wooded lands, into the hands of Robert Boyle. The latter also opened new works further inland at Araglyn, Kilmacow, Lismore, Cappoquin, and Lisfinny. Close on Boyle's heels came the Coote family, but their works were far inland, in Leitrim, Cavan, Donegal, Leix, and Roscommon. Because of their remoteness many of these works were burnt in the 1641 rebellion. At the same time as the Cootes the Wandesford family was setting up works in Carlow and Kilkenny, utilising the wooded valleys of the Nore and Barrow.

In the 1640s the woods in the Wicklow valleys were exploited for fuel, and some of the ironworks using this supply were in production for over a century. Before 1640 charcoal had been exported from Wicklow to south Wales despite the difficulties in transporting it. In 1640 an Englishman named Bacon (possibly a relative of Sir Francis Bacon), who was interested in ironworks in the Forest of Dean, came to Wicklow and founded a dynasty of ironmasters. Bacon imported ore from south Wales and used local charcoal to smelt it. On many valley sides the round circles formed by his charcoal hearths can still be seen. Bacon's only child, a girl, married a Chamney, and this family carried on ironmaking in Wicklow until after 1760. They are said to have controlled over fifty works of various types.

The last wooded area to be exploited by ironmasters was the remote south-west, where the timber in the isolated valleys was too difficult to get at to have any great commercial value in itself. Here the Pettys, Whites, and Brewsters opened works in the late seventeenth and early eighteenth centuries. By the middle of the eighteenth century the ironworks had become a byword for the inroads they had made into the oak and arbutus woods.

The point has been put forward that the Irish ironworks were not blown out because of lack of fuel but because foreign iron was cheaper and better, but contemporary observers are unanimous that they did close because local woods were destroyed. The woods could have been preserved if coppicing had been practised, but this was not done. It is extremely unlikely that ignorance of coppicing was the reason—all the known ironmasters were Englishmen and must have been familiar with coppicing. Land which in 1600 was wooded could be used for more profitable activities than ironmaking once that wood was cleared. Much of the woodland had lain not on poor marginal hill country but on lowland that was needed to support a rapidly growing population and to carry the stock that was the basis of the export trade in provisions.

(See pages 156–8 for notes)

CHAPTER 4

The Timber Trade in the Seventeenth Century

The following four chapters are concerned with the timber trade of Ireland during the seventeenth and eighteenth centuries, the price of timber, and the timber merchants of Dublin in the eighteenth century.

The seventeenth century saw a rapid expansion of the export trade but before the end of the century timber was being imported, and not all the imported timber was exotic wood. By about 1720 the direction of trade was completely reversed and imports of timber greatly exceeded exports. The volume and value of imported timber grew during the eighteenth century until by the Anglo-French wars at the end of the century imports valued nearly £200,000. As would be expected, the chief overseas suppliers in the early eighteenth century were Scandinavia and the countries that bordered the Baltic but as the century wore on the British Colonies, or Plantations as contemporaries called them, on the eastern seaboard of America became important suppliers.

Ireland was particularly dependent on imported timber as she completely lacked softwoods and she also needed to supplement her indigenous hardwoods. The trade in the export of salted provisions was bound by statute to use hardwood staves and the distillers and brewers also casked their products in hardwood.

It is clear from the amounts of standing timber offered for sale during the eighteenth century that the deforestation of Ireland in the seventeenth century was not so complete as has been supposed, and when the demands of the casking trade for wood are compared with the amount of timber imported it becomes apparent that some at least of the casks were made of local wood. It is possible that the increasing demands of the provision trade in the late seventeenth and early eighteenth centuries rather than the scarcity of local timber accounted for the virtual cessation of the export of timber after 1700.

Small quantities of timber were sent from Ireland to England during the sixteenth century and in the following century, when English control over the whole country was finally established, the amount greatly increased.

One of the inducements used to persuade settlers to come to Ireland at the end of the sixteenth century was the profits offered by the exploitation of the woods. In the document setting out 'Motives and Reasons' why the City of London should participate in the Plantation of Ulster the abundance of wood was stressed: 'All sorts of wood do afford many services for pipestaves, hogshead staves, barrel staves, clapboard staves, wainscot . . .'[1].

The export trade which developed was concerned not only with manufactured staves but also with semi-manufactured timber in the form of plank and with timber in the round. Timber and staves were shipped to Scotland, England, Holland, Spain, France, the Canary Islands, and various southern European ports. By 1615 Ireland was sending thirty cargoes of staves each year to the Mediterranean and in 1625 it was said that France and Spain casked all their wine in Irish wood. Scotland was importing wood from Northern Ireland early in the century, even though the trade was forbidden, and receiving in return pine masts for the shipyard at Coleraine.

Although small quantities of staves were produced in many parts of the country—for example, Killybegs in Donegal was exporting them to Bristol in the 1680s—the three main regions of manufacture were the Bann valley, the Slaney valley, and south Cork. This concentration of the industry is in marked contrast to the dispersal of the ironworks, which were found in nearly all wooded areas.

The reason for the concentration of stave production is found in the difficulties of transport. For timber to be profitably exploited it must be near or easily transportable to a port. Kenmare had a small trade in staves in the middle years of the century, Spanish ships from Galway calling to collect them, but the cost of transporting the timber more than a mile from Kenmare exceeded the cost of felling. Transport was always a difficulty in Kerry—the wood felled in the Glenflesk valley, which enters Lough Leane at Killarney, had to be carried on horseback to Castlemain Bay, and similarly wood in the Carragh valley had to be carried to the coast—and because of this difficulty Kerry never developed as a wood-manufacturing area and the timber was eventually used up in ironworks.

Even areas which lay round a navigable river had their difficulties: rapids on the Bann necessitated trans-shipment, and in Cork the Blackwater and its tributaries was only usable during the summer months.

The timber workers were often English. As early as 1584 it was suggested that 'immigrants . . . and other artificers of timber work' should be brought from England[2]. Raleigh's 200 timber workers on his Youghal estate were English; and when Peter Brousdon surveyed the Kenmare woods he stated that the maximum which could be cut in a single summer would be 1,000 tons and that it would be necessary to bring over English workers. Sir George Rawdon observed that in the Portarlington woods '. . . but—which I forbode to take notice of—Mr Robert Leigh is a Roman Catholic and I observed those employed in the woods etc are Irish, which some English there (under the rose) were not so well satisfied with. But this point I thought it not for me to touch upon'[3].

There were both English and Irish working in the woods on the Bann in 1611. In all, the workers numbered about 100 and included 32 fellers, 20 lath tenders, 9 sawyers, 8 wainsmen, 4 timber squarers, 4 shipwrights, 3 overseers, 15 men rafting timber down the Bann, and 9 men working the cots. Fifty men worked in the woods carting timber to the river, using 33 oxen and 3 horses[4].

Exploitation of woods started about fifteen years earlier in the south of Ireland than in the north. Raleigh was one of the first settlers to develop the trade. It was said in 1611 that, because of his influence with the Privy Council, he had shipped staves from south Cork for ten years with impunity in spite of an injunction of 1596 forbidding their export from Ireland to England. Raleigh's partner, Henry Pine, also made staves for export in Shillelagh and when he quarrelled with Raleigh the latter alleged that Pine had made over £4,000 privately from their jointly owned woods[5].

The injunction of 1596 was as difficult to enforce in the north as in the south and staves were illegally exported. Lord Deputy Chichester admitted his inability to control the trade when he wrote in 1611 to the Privy Council, 'I find it almost impossible to restrain the making and the working of the timber into pipe-staves without seizing on them when wrought and bought into the port towns which will beget much clamour and offence'. Two years later he was upbraided for permitting the cutting of staves, not only by British subjects, but also by foreigners, and he was directed to take steps to conserve the woods and prevent the export of staves[6].

The greater part of Raleigh's lands were acquired by Richard Boyle and he used their timber both in his ironworks and for stave-making. In his diaries Boyle recorded transactions involving about 4,000,000 staves (approximately 500,000 cubic ft) between 1616 and 1628.

The export of staves was again forbidden in 1615, and in an effort to enforce the prohibition a licence was granted in 1616 to Henry (or Richard?) Milton of Youghal, giving him the sole right to make and export them; but Milton abused his monopoly and it was withdrawn. One of his transactions, with Boyle in June 1618, was that the latter should pay him 6s 8d a thousand to export 71,000 staves[7].

The East India Company also owned woods in south Cork and used them both for shipbuilding and for making staves, in the export of which they 'employed many vessels'. Despite the prohibition of 1615 they were allowed to send out of the country 'the provision of timber' which they had made in Ireland for casks and staves. The Company drew on Irish woods at least until the 1640s, though not necessarily from the south Cork woods. For example, in 1636 they commissioned a Mr Stevens to treat for 100 loads of Irish timber yearly, and more if available. Four years later, in 1640, they contracted for 300,000 staves, half of them the large pipestaves, and 50,000 headings, and in 1642 John Sydney's ship, *The Dolphin,* brought 30,000 pipestaves to their English yards[8].

The second major area of stave-making was Shillelagh in the Slaney valley. As has been already pointed out staves were being made there at the end of the sixteenth century, and Chichester, commenting on the increasing activity in 1608 wrote to Lord Salisbury, 'Mittene and others have bought it from Sir Henry Harrington . . . the greatest part they intend to convert into pipestaves . . . these woods will yield sufficient store to furnish the king for his shipping and other uses for twenty years to come'[9].

By the mid-1630s most of the wealthy men in Enniscorthy were timber merchants employing over 100 men to raft the wood down the Slaney[10].

Shillelagh was of particular interest to Wentworth, who succeeded Chichester as Lord Deputy, for he had estates of his own in the county and to this day south Wicklow is sometimes referred to as 'Black Tom's country'. Yet another attempt to control the stave trade had been made in 1625 but it had met with opposition from Chichester who argued that permission should be granted to export staves to London as the ban on their export had led to bitter complaints from men who had hired ships for the purpose. For example, one Calcott Chambers, had nearly 250,000 staves left on his hands as a result of the embargo. Wentworth, however, tackled the problem with vigour and to control the trade and increase revenue put an export licence of 10s a thousand on hogshead staves and £3 a thousand on pipestaves, plus customs duty of 4s and 8s respectively. He arranged for all licences to be granted through himself and fixed the annual total to be exported at half a million[11]. But exemption from the regulations could be obtained: a Samuel Neale was allowed to send 190,000 staves and 20,000 headings (about 34,000 cubic ft) annually to London from Wexford tax free; and John Crane, a contractor for victuals to the navy, was reimbursed the duty of £145 he had to pay on 48,000 staves which he had contracted for before the levy was laid. Wentworth was also willing to exempt himself. In 1641 his agent was given permission to export Shillelagh timber 'for the private advantage' of his employer, and one of the charges laid against Wentworth at his trial was that he exploited the woods for his own advantage[12].

The restrictions on export were still in force in 1649, for in that year an Edward Lewcknow petitioned the Privy Council for permission to export 200,000 pipestaves to England. He had, he explained, lost heavily on ventures in Ireland and some of his debtors were willing to pay up to £500 in staves.

Immediately after the Restoration the Earl of Strafford was restrained from making staves in Shillelagh until it had been ascertained if the wood was suitable for shipbuilding. By 1669 he had sold several thousands of staves to Lawrence Wood of London[13]. When Brousdon reported on the timber in 1670 he said that it was unsuitable for shipbuilding, and staves continued to be made. By the end of the 1680s Wexford was exporting a third of all the staves going out of Ireland.

Page 103: *(right)* A modern state plantation of 37-year-old Sitka spruce at Knockmany Forest near Augher, county Tyrone; *(below)* young plantations, mostly of larch, at Tollymore Forest, county Down. The lower slopes of the Mourne mountains lie to the right, and in the foreground small farms occupy the shallow valley. Tollymore Forest Park attracted 180,000 visitors in 1968.

Page 104: *(above)* Camus Forest, county Londonderry, on the river Bann a few miles south of Coleraine, is a hardwood plantation used for amenity. Ugly tips of dredged material on the waterside, known as the Bann Dumps, have been transformed by planting and for long stretches the river flows between wooded banks; *(below)* Castlewellan Forest Park, county Down. Old woods, mostly of deciduous trees, make a frame for Castlewellan Lake which is an amenity area with accommodation for 700 cars. Note the rhododendrons behind the fisherman. Rhododendrons were formerly extensively planted on estates for game cover and are extremely expensive and difficult to eradicate.

The Timber Trade in the Seventeenth Century

The third area of major stave production, the Bann valley, drew on the woods of Glenconkeyne and Killetra. Timber from here was shipped not only from Coleraine, Portrush, and Londonderry, but also from Belfast via Lough Neagh.

During the preliminary discussions in January 1610 between the crown and the City of London Companies on the proposed Plantation of Londonderry the question of the ownership of the Glenconkeyne and Killetra woods arose. The crown wished to reserve them to itself but the City declared that only as lords of the soil could they ensure the conservation of the timber, which they stated they did not intend to sell. At the end of January it was agreed that the woods were to be granted to the City in perpetuity to be held of the king in common socage, and that they would be used for building and other necessary purposes in Ireland.

Two years later it was clear that staves were being manufactured in the woods and exported by Oliver Nugent, a tenant of John Rowley, one of the Irish Society's agents. Rowley was ordered by the society to recover the money illegally gained and divert it to public amenities such as bridges and roads. At the same time the Bishop of Derry complained to the Privy Council that Rowley himself had cut down 3,000 trees on the diocesan lands and exported thousands of pipestaves to Spain.

Because of this and other charges two commissioners were sent from London to inquire into the state of the Plantation, and in October 1613 they reported that Rowley had indeed made large quantities of staves and that another agent, Tristram Beresford, had made 40,000 staves. As a result of these depredations and the negligence of Godfrey, the City's wood ward, many of the best trees had been destroyed and many lay rotting on the ground. Rowley, it was also alleged, had even defrauded the City over cartage: the City had arranged with Oliver Nugent for cartage charges of 8s a load for squared timber and 10s 6d a load for timber in the round, and Nugent had charged 16s an ox load.

At the trial of the City for mismanagement of the Plantation before the Court of Star Chamber in 1635 spoliation of the woods was one of the charges. What the Companies could legally do, and did do, was to make staves to cask salmon. At the trial a cooper gave evidence that he had made 32,000 barrel staves a year for such a purpose, but he added that he had seen many pipestaves made for export to Scotland; and another cooper declared that on Beresford's orders he had made 50,000 to 70,000 pipestaves a year for seventeen years. Whether the woods were cut illegally or not the result was the same. At the Londoners' coming the woods were very thick but by the 1630s 'a man might see a mile through them' and the value of the timber so cut was estimated at £3,000[14].

Official complaints about the cutting of the Londonderry woods came hard on the heels of the Restoration, for, as it was explained, in Londonderry alone were the woods reserved to the state and the City itself was only permitted to take timber for building purposes. Yet, Colonel George, the overseer of the woods, gave so many warrants for timber repairs that 'there is more timber consumed in this year (1660) than would have built the city of Derry and Coleraine and this timber sold into Scotland for hogshead barrel staves to be sent to France and other parts and not the fortieth part employed for the use intended'.

TABLE 4

EXPORTS FROM BELFAST AND COLERAINE, 1683–95

	Belfast 1683–6	1686–95	Total
Staves (hds)	3,497	6,863	10,360
Plank (ft)	16,860	21,580	38,440
Timber (tons)	155	136	291
	Coleraine*		
Staves (hds)	1,537	2,197	3,734
Plank (ft)	17,160	51,450	68,610
Timber (tons)	338	800	1,138

* Excluding April 1693 to December 1694

The export of timber from the north of Ireland continued until the end of the century and the amounts exported between

1683 and 1695 are given in the preceding table. As can be seen, Belfast was exporting the greater number of staves while Coleraine was handling the bulk of the plank and timber in the round[15].

TABLE 5

EXPORTS TO BRISTOL FROM IRELAND

	1613	1655	1685
Pipestaves	71 (hds)		
Barrel staves	271 (hds)	440 (hds)	52 (hds)
Hogshead staves	190 (hds)		
Timber (tons)	4	¾	16

Figures for the whole of Ireland are available for 1635–41, 1665, 1668, and 1683–6. Exports to England alone are also available for 1691 and 1696. These sets of statistics are given in Table 6. Table 5 shows timber sent to Bristol in 1613, 1655, and 1685[16].

TABLE 6

TIMBER EXPORTS, 1635–91[17]

	1635–40*	1641	1665	1669	1683–6*	1691
Staves (hds)						
Pipe	3,589	1,206	367	1,017		
Hogshead	6,266	5,525	1,915	2,342		
Barrel		7,842	2,211	5,283		
Total	9,855	14,573	4,493	8,642	5,182	8,667
Timber (tons)		384	191	660	752	
Plank (tons)		209		159		

* Yearly average

The effect of the Anglo-Dutch war on the trade is evident in the abnormally low figures for 1665, but it would appear that the trade in staves had reached its zenith before the middle of the century, though the export of timber in the round increased. The value of the export trade in wood was small and it only formed an exceedingly small proportion of the total export trade. In 1665 the value of all timber exports was only £2,384— that is one-half per cent of the total value of Irish exports.

At the beginning of the seventeenth century the timber trade was directed towards the continent—chiefly France and Spain, but by the end of the century the continental markets were unimportant and England and Scotland were the most important customers. This change in markets was apparent by 1665, when more than twice the number of staves were sent to England than to the continent. By 1686 England and Scotland received more than five times the number of staves sent abroad. The amount of timber in the round exported to England in 1686 was more than three times that exported twenty years earlier and she was also now taking plank from Ireland.

TABLE 7

DESTINATION OF TIMBER EXPORTS, 1655 and 1683–6*

	Staves (hds)		Timber (tons)		Plank (ft)	
	1665	1686	1665	1686	1665	1686
England	3,119	3,285	191	752		10,930
Scotland		1,109				
France	1,372	447				
Spain		390				

* Yearly average

The size of the timber trade carried on by the more important Irish ports is shown in Table 8, which contains the totals exported from 1683 to 1686 inclusive.

The trade with Scotland was entirely in the hands of Londonderry, Coleraine, and Belfast; England was supplied from the south-eastern ports, particularly Wexford, and the continental trade was carried on from the south-west and west ports. By the 1680s Wexford was by far the most important timber port, handling a third of the staves and timber and a quarter of the plank. Waterford, which in 1611 was the second largest timber port in Ireland, had become the least important timber port by the end of the century, and Ross, described in 1611 as 'a poor ruined town', could show a larger trade in staves than either Dublin, Cork or Sligo.

So far timber has been considered as an export commodity,

The Timber Trade in the Seventeenth Century

but the timber trade was also closely connected with the export provision trade in salted beef, butter, pork, tallow, and fish to British Colonies in North America. This trade dated back to the beginning of the seventeenth century, but it expanded very rapidly during the 1660s when the export of live cattle to England was prohibited in the interests of English graziers. The effect of this embargo on the Irish economy was disastrous—the numbers of cattle shipped to Britain fell from 60,000 in 1660 to 1,454 in 1669. Denied an outlet for live cattle the Irish turned to the provision trade and the extent to which it expanded between 1641 and 1685 is shown in Table 9.

TABLE 8

TRADE OF CHIEF TIMBER PORTS
(Totals 1683–6 inclusive)

	Staves (hds)	Timber (tons)	Plank (ft)
Baltimore	1,958	169	700
Belfast	3,497	155	8,160
Coleraine	1,537	338	17,160
Cork	266	56	
Dingle	1,032	107	5,800
Dublin	663	808	800
Kinsale	421		
Limerick	1,466	214	100
Ross	760	150	
Sligo	665	18	
Waterford	133	22	
Wexford	7,626	993	10,980
Other ports	705	75	
Total	20,729	3,105	43,700

TABLE 9

THE PROVISION TRADE, 1641–85
(Number of barrels exported)

Commodity	1641	1665	1669	1683	1685
Beef	15,200	29,200	51,800	79,240	72,200
Pork		1,250	770	594	2,510
Butter	17,410	13,200	29,000	68,510	134,700
Cheese		6	159	615	2,043
Tallow	10,050	10,500	19,100	18,990	20,700
Fish	46,610	23,620	30,940	13,900	

In order to protect and regularise the trade it was enacted in 1698 that provisions for export were to be casked in barrels of 'sound, dry and well seasoned timber' which were to weigh 40lb[18]. It is reasonable to assume that before 1698 the weight of a cask was about 40lb, and indeed may have been more, as two of the reasons behind the act were (1) to prevent the use of unseasoned timber and (2) to ensure that the cask was not so heavy as to reduce the weight of the contents, which were standardised at 2cwt.

The volume of timber used by the provision trade, together with the volume exported as staves, plank, and timber in the round is given in Table 10. It can be seen that while the export of timber decreased the volume of timber used by the export provision trade increased, and the total volume of timber leaving the country remained roughly constant. As little timber was imported into Ireland during the seventeenth century it is reasonable to assume that the decrease in the amount exported was due not so much to shortage but because the provision trade was absorbing the local supplies.

TABLE 10

VOLUME (CUBIC FEET) EXPORTED, 1635-91

	Timber trade	Provision trade	Total
1635-40[a]	172,700		
1641	199,300	59,100	258,400
1665	66,500	54,500	121,000
1669	151,700	92,600	254,300
1683	108,700	124,900	233,600
1686	101,100	161,000	263,100
1697[b]	13,500		
1711[c]	4,100	159,600	163,700
1717[c]	870	204,500	205,100

[a] Yearly average [b] to England only [c] excludes plank

Although the general flow of timber was out of Ireland during the seventeenth century, a certain amount of softwoods, which were not native, were imported. By the middle of the century the use of deal for floors was becoming popular and small quantities—388 hundreds in 1665—were imported from Scandinavia. No deliberate seasoning took place in their country of origin, though a certain amount occurred during their passage down the rivers to the sea and on their journey to the British Isles. Thus it was necessary to complete the seasoning, and one finds Sir George Rawdon remarking in 1670, 'Deals are very cheap at Belfast and I am ordering a large quantity to have them seasoned'.

TABLE 11

TIMBER IMPORTS, 1683–6

	Deals (hds)	Timber (tons)	Plank (ft)	Wainscot (pieces)
1683	3,036	347	4,580	160
1684	2,484	283		
1685	6,084	305		
1686	1,734	277		

The amount of foreign wood imported into Ireland between 1683 and 1686 is given in Table 11. Scandinavia was the chief source of the deals; at least two-thirds of the timber came from Scotland; half the plank was walnut from France; and the remainder of the plank came from Scandinavia and North America.

(See pages 158–9 for notes)

CHAPTER 5

The Timber Trade in the Eighteenth Century

The eighteenth century saw a reversal of the direction of the trade in timber: from being primarily timber-exporting Ireland changed to a timber-importing country. The change was beginning at the end of the seventeenth century and by 1711 it was well established, for in that year imports totalled £11,000 and exports £600.

The value of timber sent out of Ireland between 1711 and 1760 is shown in Table 12[1] and it can be seen that it fell to well under £350 annually after 1730.

Even these figures, small as they are, give an exaggerated picture of the amount of Irish wood exported, for deals, masts and spars were brought into the country from the Baltic and then exported to England.

Ireland had no indigenous conifers and the dangerous dependence of the country on outside supplies of softwood was noted by a writer in 1729. 'Our imports from thence (the Baltic)', he wrote, 'consist principally of deal boards, timber of all sorts . . . which we cannot do without . . . As for our importation of

The Timber Trade in the Eighteenth Century 113

wood, I am afraid we shall not in a long time, if ever, save in that article, even should we plant, to which we seem generally to have so great a disinclination, for as we increase and improve our demands for it will still be increasing, and when Norway and the Baltic fail, we must look out for another market to buy at a greater expense'[2].

TABLE 12

TIMBER EXPORTS, 1711-60 (VALUED IN £)

	Deals	Other timber	Total
1711	—	620	620
1717	270	15	285
1719	157	820	977
1725	167	407	574
1730	232	294	526
1732	91	143	234
1740	17	257	274
1745	150	70	220
1751	11	90	101
1756	27	98	125
1760	93	245	338

The value of imported timber is given in Table 13. The amount spent increased from £11,000 in 1711 to £187,000 in the early 1790s.

TABLE 13

TIMBER IMPORTS, 1711-93
(Average yearly value in £)

1711	11,300
1719-27	46,000
1733-6	34,100
1744	31,700
1751	56,000
1761	49,000
1764-9	103,000
1770-9	105,600
1780-9	131,100
1790-3	187,400

The steep rise in the value of imported timber during the 1760s was chiefly due to increases in the amount of deals, staves, and raw timber. The greatest increase was shown in deals, whose value rose from £22,300 to £44,000, and in timber, whose value increased from £12,300 to £26,100.

There was a general increase in all types of imported goods during the eighteenth century as well as timber. Taking two examples, coal and sugar, the value of imported coal between 1720 and 1780 rose from £46,000 to £165,000, and of sugar from £53,000 to £360,000. Whereas imported timber represented 5 per cent of all imports during the 1720s, by the 1790s it formed 3½ per cent.

Comparison may be made with the quantities and value of timber imported into England during the eighteenth century. In 1721 England imported 4,000,000 deals and 29,000 tons of softwood costing £74,000; the Irish imports of deals and softwood from the Baltic cost £35,000. By 1790 English imports from the Baltic comprised 5,750,000 deals and 266,000 tons of wood costing £755,000. Irish imports, comprising 2,000,000 deals and 14,300 tons of softwood, were £107,400. In other words at the beginning of the century England was importing twice as much as Ireland and by the end of the century seven times as much.

It can be seen from Table 14 that the Baltic became an increasingly important supplier of timber as the century progressed, until by 1790 four-fifths of all wood came from that area (Norway and Russia). The Baltic supplied practically all the deals, balk, and raw timber.

TABLE 14

ORIGIN OF IMPORTED TIMBER
(As percentage of total value)

	1711	1719–27	1756	1764	1793
Great Britain	11	13	7	7	4
Baltic	57	76	77	76	80
America	—	—	10	7	7
Elsewhere	32	11	6	10	9

During the first part of the eighteenth century most of the staves imported into Ireland came from Spain, Portugal, France, and Holland, but by 1750 these countries had been superseded

The Timber Trade in the Eighteenth Century

by the Plantations, which by 1790 supplied 80 per cent of all imported staves. There was a short-lived revival of Low Country trade between 1778 and 1782, when Ireland received 40,000 staves.

The pattern, therefore, was that the Baltic supplied softwood, together with a small quantity of hardwood, and the Plantations hardwood, chiefly in the form of staves.

The annual value of various types of imported timber is shown in Table 15. Deals formed the biggest single item throughout the eighteenth century but their share fell from 50 per cent in 1711 to 40 per cent in 1790. On the other hand unworked timber rose from 2 per cent to 30 per cent.

The value of the less important items is shown in Table 16.

TABLE 15
ANNUAL VALUE OF IMPORTED TIMBER IN £

	Deals	Timber	Wooden ware	Balk	Staves	Hoops
1711	5,600	200	800	300	2,700	1,340
1733–6	13,700	4,200	1,150	5,000	4,800	3,100
1740–4	15,900	3,800	3,400	2,900	2,000	1,700
1751	23,100	11,900	2,740	4,200	4,500	1,340
1761	22,300	12,300	1,120	4,300	1,760	3,800
1764–9	44,500	26,100	6,200	5,500	8,100	4,200
1770–9	42,500	30,200	6,000	5,100	8,100	8,600
1780–9	49,300	38,000	10,600	7,400	8,900	10,200
1790–3	74,300	57,900	17,400	14,500	8,800	6,900

TABLE 16
VALUE OF LESSER TIMBER IMPORTS IN £

	Canes	Clapboard	Plank	Casks	Clap-holt	Ships' auxiliaries
1711		110	220			
1733–6		420	1,130		630	
1740–4		390	1,450		210	
1751	920	1,000	3,200	380	90	340
1761	410	340	1,880	410	340	190
1764–9	850	760	5,500	180	370	540
1770–9	1,000	700	1,960	80	980	390
1780–9	1,730	740	2,800	360	410	610
1790–3	4,300	1,100	1,450	140	110	530

These include auxiliary ships' parts (oars, masts, spars), canes, empty casks, clapboard, clapholts and plank.

Table 17 shows the actual quantities of the main types of timber imports; hoops are in thousands, that is 1,200 individual pieces.

TABLE 17

AVERAGE ANNUAL QUANTITIES OF IMPORTS

	Deals (hds)	Timber (tons)	Balk (hds)	Staves (hds)	Hoops (mille)
1711	4,500	100	100	416	940
1733-6	6,600	2,080	300	32,900	18,600
1740-4	7,600	1,520	180	18,900	
1751	5,800	4,800	208	29,900	1,864
1761	5,300	4,500	214	11,700	1,900
1765-9	12,600	9,500	330	38,900	20,900
1770-9	11,900	11,000	300	38,800	43,400
1780-9	15,000	13,800	510	43,300	49,800
1790-9	19,400	21,000	760	45,700	38,700

An interesting sideline was the amount of wooden ware brought into Ireland. Wooden ware was the term applied to manufactured articles of wood, excluding coaches. During the early and middle years of the eighteenth century Great Britain was the chief supplier, but by the 1760s America was supplying about a fifth and by 1790 just over a half—also by that date she was supplying over half the imported plank. Coaches were also imported; they cost between £30 and £40 each and their value rose from £790 in 1736 to £3,420 in 1783.

Wood from the Baltic and America might be described as utilitarian timber, but there was also a small trade in luxury timber. Walnut was brought in from France before the end of the seventeenth century and during the eighteenth century mahogany appeared. Mahogany and other particular types of wood are not generally listed separately in the Irish customs returns, except for the years 1794-6 when a total of 3,690 tons was recorded. However, some mahogany was sent from England to Ireland—about a third of a ton in 1725, 19 tons in 1730, 10½ tons in 1735, and over 200 tons in 1750. The country of origin is not known, but in any case mahogany is a general term covering several different species of tropical hardwood. Dublin timber merchants described mahogany they had for sale during the 1760s as being Honduras mahogany[3].

Two other exotics appear in the 1732 customs returns: 1¼ tons of ebony and 6cwt of cocoa wood.

Comparatively small amounts of ash, beech and oak were also brought from England: 232 tons of ash in 1717 and 224 tons two years later; £960 worth of oak in 1730 and £98 worth in 1756.

A minor item was wainscot; Holland was the chief source of supply but some came from the Baltic and from North America. Until the 1760s the yearly average value was under £200, but later it fluctuated between £400 and £600. Wainscot was made of either hard or soft wood; Norwegian oak wainscot fetched 6s to 7s a square yd and deal wainscot 3s to 3s 6d[4]. As a house of any pretensions in the eighteenth century had wainscot in some at least of the rooms, and as it was during the second half of the eighteenth century that so many of the numerous Georgian houses in Dublin were built, it must have also been manufactured in Ireland.

Practically every port in Ireland handled timber, but as can be seen from Table 18 about two-thirds was unloaded at Dublin, Cork, and Londonderry. Cork, the centre of the provision trade, received nearly half the staves, but Dublin, the capital and largest town in the country, took a quarter of the staves and over a third of the balk and deals.

How dependent was Ireland on foreign timber during the eighteenth century? It is difficult in the twentieth century to appreciate the large part wood played in a society which lacked cellophane and plastic containers, unlimited numbers of paper bags, 'cheap tin trays', cheap enamelled ware, and reinforced concrete and steel girders. Raffia bags and cane baskets were used for dry commodities, but liquids were for the most part kept in wooden buckets, casks and mugs. Obviously it is not possible to estimate this domestic consumption of timber. Likewise, how far wood was used as a domestic fuel is problematical. Imported coal was used in the coastal towns and in inland regions coal and wood were supplemented by peat.

The numerous ironworks, except for the foundries at Newry, Belfast, Dublin and Cork, were completely dependent on timber for fuel, and shipbuilding made some demands on standing timber. No wooden houses were built in the eighteenth century, but floors, rafters, doors, and window-frames were of wood.

TABLE 18

PORTS' PERCENTAGE OF IMPORTS, BY VALUE, 1773

Port	Staves	Balk	Timber	Deals
Belfast	4	—	5	6
Cork	47	20	8	5
Drogheda	1	—	4	6
Dublin	12	37	24	37
Londonderry	14	—	4	4
Limerick	2	1	12	7
Newry	5	—	15	13
Ross	1	14	3	—
Waterford	5	—	17	4
Elsewhere	9	29	8	18

From the number of advertisements for the sale of wood in contemporary newspapers it is clear that the denudation of the country's timber was not so complete as has been generally supposed. The legend of the treeless Ireland may have been built on Arthur Young's statement in the 1770s that 'the greatest part of the kingdom exhibits a naked, bleak, dreary view for want of wood'. It is possible that he was referring to the very large stretches of bog that occur nearly everywhere in Ireland.

Perhaps the problem was not a complete lack of timber but its availability. Roads were often primitive and the movement of timber any distance was a costly business. The building of canals in the middle and latter part of the eighteenth century eased this problem within limited areas, and those wishing to sell standing timber near a canal did not neglect to point out that the wood could be moved with relative ease; but it is significant that advertisements for timber in some of the more remote areas continued to appear for as long as three years.

Another factor favouring imports was cost: although the cost of transport from the Baltic exceeded the original cost of the timber it was cheaper to import sawn deals than to cut them in Ireland, because pit-sawing was still used at home whereas the Scandinavians used mechanical saws driven by wind or water.

One can calculate the amount of timber needed for the export provision trade and also that needed by brewers and distillers. But though the total needs of brewers and distillers can be estimated it does not accurately reflect their consumption, for the effective life of a cask was ten years, and casks which brought wine and brandy from the continent could be used in the home industry. Distillers required on an average 290 tons of hardwood annually from 1720 to 1750, 780 tons annually from 1750 to 1780, and 3,300 tons annually from 1780 to 1800. Brewers' needs were greater than those of distillers: from 1750 to 1792 they used an average of 25,100 tons of wood a year. These figures, both for distillers and for brewers, are based on official returns and take no account of the vast quantities of illicit spirits and beer on which duty was not paid.

The average annual weight of wood in casks bringing in wine and brandy from abroad came to 1,200 tons from 1720 to 1750 and 4,700 tons from 1750 to 1780. If these figures are deducted from our previous estimates, we find that the Irish manufacturers needed about 20,000 tons of wood annually above the weight of wood in imported casks.

The casks in which beef, pork, and butter were sent abroad were non-returnable and one is on firm ground in saying that each provision cask represented 40lb of wood. Between 1700 and 1750 the provision trade used 4,500 tons of wood a year. After 1750, at which date the consumption first reached an annual average of 5,000 tons, until the end of the century the yearly average came to 7,000 tons.

The third trade dependent on wooden containers was the dry provision trade in corn (oats), wheat, meal, and flour, but calculating the amount of timber used in this trade is difficult. Containers could be used more than once; in certain years there were both imports and exports of the same commodity; the size of a 'barrel' varied in different parts of the country and for different cereals; and even if the size of a 'barrel' was known no regulation governed its weight such as those governing the meat containers. However, a minimum figure can be arrived at for the latter part of the century. Between 1772 and 1792 the average yearly excess of cereals exported over those imported was 74,300 barrels. By that time the average barrel contained about 2cwt, so if it weighed the same as a provision cask then it took 1,330 tons of wood a year to barrel the surplus of export over import.

Summarising these estimates it can be postulated that towards the end of the eighteenth century distillers, brewers, export provision merchants, and cereal merchants needed around 30,000 tons of wood a year above such casks as came in from abroad.

Unfortunately the customs returns for timber do not differentiate between the sizes of staves, and as the stave for a provision barrel was very much smaller than a stave for a pipe the weight of staves cannot be calculated. Even if all the staves imported were of pipe size in 1780 their weight would have come to 24,000 tons, which is less than that used by our three chosen consumers; and even if these three trades had also taken all the raw timber imported in 1780—13,800 tons—that would only have left a surplus of just under 8,000 tons for all other purposes.

The question of how far imports met the demand can also be looked at from a financial angle. The provision trade used an annual average of 380,000 casks between 1764 and 1793. The wood in each cask cost 1s 6d so the value of wood in them was about £28,500. The value of the 74,300 casks allotted to the cereal trade was £5,600. The 20,000 tons allotted to brewers and distillers represents about £80,000 worth of raw timber. These statistics give a total of £114,100.

Reference to Table 19 shows that the value of staves and raw timber imported was considerably less than this.

TABLE 19

ANNUAL AVERAGE VALUE OF IMPORTED TIMBER IN £

	1764–9	1770–9	1780–9	1790–3
Staves	8,130	8,080	8,940	8,790
Hoops	4,180	8,620	10,160	6,940
Clapholts	760	700	740	1,110
Timber	26,120	30,220	38,030	57,850
Total	39,190	47,620	57,870	74,690

One last item remains to be mentioned—hoops. Casks were hooped with either hazel or osier (willow) and after 1698 each cask for provisions had to be bound with twelve hoops (this was increased to sixteen for herring barrels in 1733). From 1761 to 1800 the annual value of hoops used by the provision trade alone was £9,000. Reference to Table 19 will show that the hoops imported were not on the whole sufficient to meet the needs of this one trade. From this it can be said that a large proportion of the hazel and willow used in the manufacture of hoops must have been made from local wood.

(See page 159 for notes)

CHAPTER 6

The Price of Timber and Timberworkers' Wages

As would be expected the price of timber rose during the two centuries under review. There were three main items involved in the cost of local timber: the value of the standing tree, the cost of labour in hewing, felling and sawing, and the cost of transport. While imported timber was treated on the timber merchant's premises local timber was prepared on the spot and the finished product moved to where it was required.

The price of a standing oak in Cork at the beginning of the seventeenth century was 1s. A little over a century later, in 1731, 2,150 oaks in Shillelagh were valued at £3 17s 0d each and by 1780 the few remaining trees were worth £13 10s 0d each.[1]

Wood was also sold by the acre. Two examples may be given: 1,300 acres of wood on the Brownlow estate near Lurgan fetched £7 an acre at the beginning of the eighteenth century, and the Cappanellan wood of 100 acres near Kilkenny fetched £80 an acre at the end of the eighteenth century. These figures, however, tell us nothing about the value of an individual tree for the age of the woods is not known. If they were full grown there would have been between thirty and forty to the acre, but if they were younger there would have been anything up to 1,000.

On the whole, until the Restoration, felled and squared timber fetched about 8s a ton; and unsquared raw timber a little over half that price. A breakdown of the cost of a ton of timber is provided in Peter Brousdon's reports on available wood in different parts of Ireland in 1670. These show that labour and transport cost considerably more than the raw material.

TABLE 20

COST OF TIMBER AND TRANSPORT, 1670[2]

Kenmare	
Timber, per ton	5s
Hewing	3s
Transport to river (1–2½ miles)	5s
Sea carriage to England	30s
Shillelagh	
Timber, per ton	8s 6d
Felling and squaring	2s 6d
Sawing in plank, per 100ft	3s
Land carriage to Enniscorthy (10 miles)	10s
Enniscorthy to Wexford by boat (10 miles)	2s
Sea carriage to England	28s
Coleraine	
Compass timber*, per ton	1s
Felling and hewing	2s 6d
Land carriage to Bann, drawing over rapids	3s 9d
Water carriage to Coleraine	1s 8d
Land carriage from Coleraine to Derry or Portrush	2s 1d
Making 2 and 3ft treenails, per thousand	2s 6d
Transport to vessel, per thousand	4s
Total cost of treenails, per thousand	30s
Felling, sawing and squaring plank, per 100ft	20s
Transport of plank, per 100ft	20s

* Branches, not the main trunk

Brousdon explained that the timber could not be shipped from Coleraine in the naval vessels specially used for transporting timber as the Bann estuary did not take ships drawing more than 8ft of water. His figures well illustrate the large fraction of the total cost of timber that transport represented, why landlords in remote areas sold their woods to ironmasters, and why ironmasters found it uneconomic to bring timber any distance to their ironworks.

In England during the seventeenth and eighteenth centuries it was generally reckoned that the cost of transporting oak any distance was equal to the cost of the timber and a haul of over twenty miles raised the price to an uneconomic level. Another factor which operated in both England and Ireland, where pit-sawing was the rule, was the cost of labour.

The Dutch War of the post-Restoration period caused the price of imported timber to rise to £2 5s od a ton in Dublin and to £1 15s od–£2 a ton in Belfast[3]. There was a general rise in the price of native timber during the 1720s. Imported yew had fetched 12s a ton and imported ash 11s a ton in 1717. By 1730 local Kenmare wood was 12s a ton and in Dublin in the 1750s the average price of Irish wood was £2 15s od. Imported exotic hardwoods were of course more expensive: walnut was £7 a ton and mahogany £10. The small cargoes of ebony and cocoa wood brought into Dublin yielded £21 a ton for the former and £11 14s od a ton for the latter[4].

A further increase in prices took place at the end of the century during the Anglo-French war. Imported Baltic timber rose to £4 10s od a ton and American to £6 16s od. The price of local wood varied according to species: birch and ash fetched £5, oak £5 10s od and elm and beech £6 10s od.

It now remains to consider the price of manufactured timber, that is of staves, deals, plank, and clapboards. At the beginning of the seventeenth century staves were priced at between £4 and £8 the thousand according to their size. This was increased to £10 for pipestaves during the Anglo-Dutch war, but by the end of the century the price had fallen back to what it had been at the beginning.

Imported staves at the beginning of the eighteenth century cost £13 a thousand for pipestaves and £6 10s od for the smaller barrel staves, but prices fell during the century until by 1790 the average cost was £2 10s od a thousand—though this average includes the price of the very small staves used in provision casks.

All deals used in Ireland were imported and by the end of the seventeenth century the average price was £3 10s 0d the hundred. By the 1750s Norwegian deals were £1 10s 0d a hundred, but spruce deals fetched betwen £6 and £7. At the beginning of the 1790s the average price was £4 5s 0d.

Clapboard cost about £3 5s 0d a hundred in 1720 and the price rose to £6 5s 0d in 1750, at which level it stayed until the 1790s.

In both England and Ireland timber was sawn by hand. Sawmills operated by wind or by water had been introduced on the continent during the fifteenth century, but an effort to introduce mechanical sawing near London in 1663 was frustrated by a mob of artisans who feared for their livelihood and in 1768 a similar mob destroyed a sawmill set up near Limehouse[5].

Two proposals to set up sawmills in Ireland were made during the seventeenth century, though whether anything came of them is not known. The first was in 1616 when a planter, Richard Rollston, who had land in Armagh, applied for and received permission to set up 'sawmills . . . going with wind or water . . . in any place or places in Ireland . . . for the cutting of all sorts of boards and timber with a prohibition of all others to do the same'. The second was made at the Restoration when Charles II wrote to the Lord Lieutenant, 'We hear that Sir Hugh (Mydeton) is anxious to set up windmills in several places for the more speedy, easy and cheaper sawing of timber and boards, a thing not formerly used or known among our people in Ireland. As the setting up and erecting of such mills will cost much money, we authorise you to grant Sir Hugh Mydeton by patent the sole right to set up and use such mills for fourteen years in such places as he shall think convenient'[6]. Pit sawing was so expensive that it was cheaper to import sawn deals from the Baltic than to cut them at home, even though the cost of transport from the Baltic was greater than the cost of the timber.

Timberworkers during the seventeenth century were paid by the piece rate rather than by the day or week. Richard Boyle paid £1 12s 0d a thousand for the making of staves and the timber merchants at Enniscorthy paid £1 10s 0d. Fellers and squarers received between 2s 6d and 3s a ton and sawyers 3s per 100ft.

In the eighteenth century workers on estates were generally paid by the day or week, whereas in the towns the timber merchants paid a piece rate. Sawyers and carpenters on the Kenmare estate received 1s a day during the 1720s, and during the 1770s coopers got £7 a year and carpenters £12. On the Carew estate in Waterford coopers earned 1s 1d to 1s 6d a day during the 1770s and on the Headford estate in Meath carpenters got 1s a day in 1786. At the end of the century Kilkenny fellers were earning 1s 6d to 2s a day as compared with the daily wage of 8d to 10d paid to labourers.

In the north of Ireland the cost of labour was between a third and a half of the value of an article in 1800. A cart, plough, and harrow cost £1, 6s and 4s 6d respectively, of which 7s 6d, 3s and 2s were wages for the work done[7].

A small sidelight on industrial unrest in Ireland in the eighteenth century is provided by the grievances of the timberworkers. In 1776 some Dublin timber merchants advertised for sawyers and lath-splitters from any part of the British Isles to work in deal yards; and their advertisement was immediately countered by a declaration in the press by Dublin sawyers that there was no shortage of workers in the city, that most of the employers paid 1s 3d a dozen for deal cutting, and that the merchants who were advertising for new workers only wanted to pay 1s a dozen and had dismissed employees 'without due cause'[8]. The coopers of Limerick complained in 1769 of a reduction in piece work rates. They demanded a return to the prices of 1765, which even then had been 'two pence in each piece of work under the prices of Dublin, Cork and Waterford'.

(See page 159 for notes)

CHAPTER 7

The Timber Merchants

There were at least 170 timber merchants in Dublin between 1750 and 1800[1]. The number in any one year increased from eight in 1751 to thirty-one in 1765 to forty-nine in 1799, an increase in line both with the physical expansion of Dublin and with the growth of imports of timber.

Six of the timber merchants were women who continued to run the business after their husbands died, and twelve were sons who represented a second generation in the firm. One of the longest active merchants was Bridget Moss, who took over when her husband died in 1758 and ran the timber yard in Golden Lane until 1786.

Some of the timber merchants combined the sale of timber with an allied occupation, the most common combination being timber merchant and carpenter; of the forty-four timber merchants in Dublin in 1793 ten were also carpenters. There were others who were also architects, builders, coopers, paper-makers, cabinet-makers, turners, timber measurers, slaters, joiners, pump borers, and valuers and improvers of wood. One described himself as a deal merchant, another as a mahogany merchant, and a third as a hardwood merchant. A fourth,

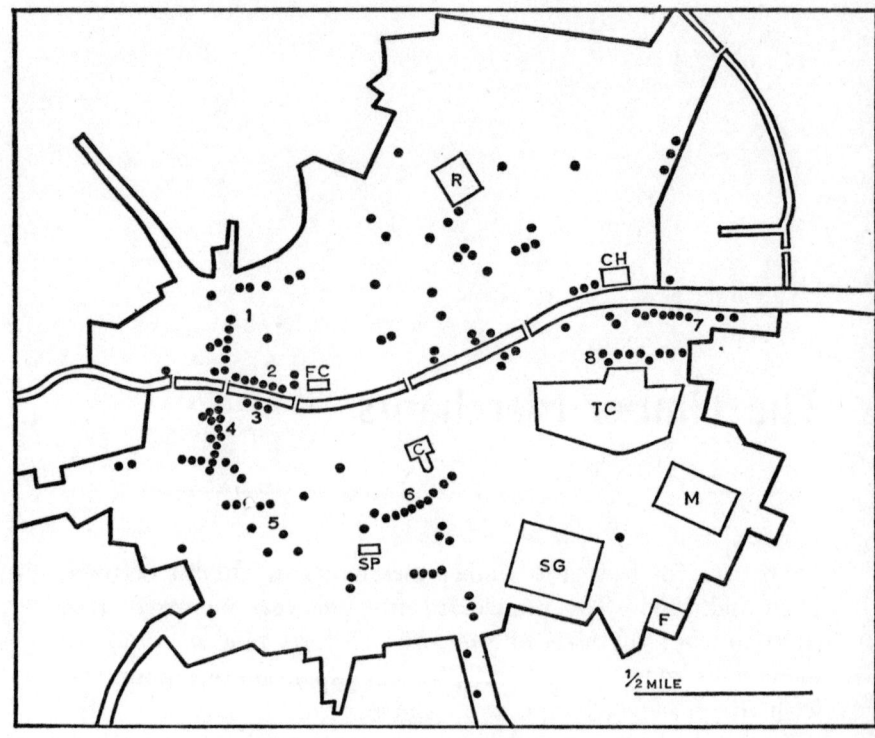

Map 6. Dublin timber merchants, 1750–1800

R	Rutland Square	1	Smithfield
CH	Customs House	2	Arran quay
F	Fitzwilliam Square	3	Ussher's quay
C	Dublin Castle	4	Bridgefoot street
SP	St Patrick's Cathedral	5	The Coombe
SG	St Stephen's Green	6	Golden Lane
M	Merrion Square	7	City quay
TC	Trinity College	8	Townsend Street

Thomas Hawkshaw, was the Agent and Receiver for the Blue Coat Hospital.

The timber merchants tended to concentrate along the quays and in the older parts of the town (Map 6). Those along the Liffey were found at Ussher's and Arran quays, near the Four Courts, and at City quay, opposite the Customs House. Away from the river and on the north side the majority were found in the streets round Smithfield. On the south side of the river they were found in Bridgefoot street, which ran up from Ussher's quay; in the Coombe; in Golden lane, at the rear of Dublin Castle; and in Townsend street, along the north side of Trinity College. As would be expected, they were virtually absent from the newer and more select residential areas that lay on the east side of Dublin north and south of the river. Like other merchants of the time they lived in houses attached to their yards; and these yards could be quite large, one at the North Wall having a frontage of 44ft and a depth of 500ft.

Timber measurers, as opposed to timber merchants who were also measurers, were few, probably under a dozen. One of the more eminent was Levi Hodgson of 5 Poolbeg street; he was the son of Daniel Hodgson, who had been appointed City Measurer by the Guild of Merchants in 1753 and held the post for 45 years. In 1777 Levi produced *The complete measurer* (later reissued as *The modern measurer*) expressly for the guidance of Irish timber merchants; and this book, based in part on Hoppus' earlier work, was such a success that by 1801 it had run to ten editions. It contained elaborate tables showing the costs of different sizes of plank and the various timber items used in house construction; and advice on measuring wood when buying or selling. The most important point to note when buying wood, according to Levi Hodgson, was to see that bark, sap, and defects such as knots, shakes, etc were allowed for except when measuring for freight. When measuring timber 'fresh off the water' the measuring line should be drawn tightly on the piece and laid slack on the rule. Measuring with callipers led to errors unless the piece was a perfect square, and in measuring knee timber each arm should be measured separately. To girth regular timber it should be measured at the centre, but if it bulged it should be measured at the spot inclined to the butt end or the average of two or more separate girths should be taken. With flitch-shaped Baltic timber the buyer should not girth it but should take the breadth and depth, and with mahogany allowance should be made for the sap of the waves and the keft of the saw.

Apart from the day-to-day sales in the timber yards professional auctioneers sold off whole cargoes of timber at the quayside, and on occasions the entire stock of a timber merchant was sold on his retirement or death. The leading firm of auctioneers was John Kirchoffer & Joseph Graves. Their conditions of sale varied slightly from time to time, but in general credit was given, on security, for from three to twelve months, or 8 per cent discount was given for cash. One guinea was demanded on deposit for each lot sold. At any one sale a purchase of under £10 had to be paid for immediately, sales of between £10 and £20 were allowed three months' credit, and six months' credit was given on sales over £20. Buyers not removing their timber within two weeks forfeited their guinea deposit.

Kirchoffer & Graves sold the more common species by the ton: American oak was put up in 4 ton lots and Norwegian deals in 10 ton lots. On the other hand Honduras mahogany was sold in lots of 1,000ft.

Lesser known firms auctioned timber at smaller ports, and often could not afford to allow such credit terms as the Dublin firms. For example, E. Hardman & Son, auctioning timber at Dublin Gate in Drogheda, demanded 1 per cent at the time of sale, and the rest before the timber was removed, which was to be within ten days.

During the 1760s two timber merchants' stock was sold off by auction in Dublin. The first, in 1761, was on the death of Robert Ball, who had a yard near Ship Buildings on the North Strand. At this auction half the money had to be paid on the day of sale and the rest within three months on approved security. Ready money gained 5 per cent discount. At the second auction, in 1764, on the retirement of Leonard Buckley of Little Britain street, three to six months' credit was given on purchases of over £5 if security was given.

While conditions of sale in Dublin were more or less standard, selling in the countryside was done by the owner of standing timber, or his agent, and conditions of sale tended to differ considerably.

The sales were advertised and various information given. The distance to any nearby ironworks or tannery was often mentioned as a selling point, as was accessibility to water transport, particularly canals. The scarcity of timber in the area, and the uses to which the wood could be put, especially if it was gross timber suitable for mill axles, housebeams, or shipbuilding, were worth mentioning; and owners of ash tended to emphasise its suitability for coachmaking. Trees were generally sold standing; but occasionally felled timber was put on the market, and then it was usual to give the length of seasoning, which varied from one to twelve years according to thickness.

The name and address of the person who could show a prospective buyer round—usually the woodranger or owner—was given, but if, as frequently was the case, the owner was an absentee landlord then the name of a friend or agent in a nearby town was supplied.

The price expected was very rarely stated. The usual procedure was to sell either by public auction or by private treaty. In the latter case the prospective buyer was invited to send in an offer, which if he so desired would be kept secret; but he was usually warned that he must pay the postage on his communication.

Terms of sale were not often inserted in advertisements but there were a few such cases, and the following are a few random examples. It was very rare for a seller to ask for cash. The Rathmullan woods in Donegal were sold in 1741 and the buyer was asked for £200 down, £500 the following year, and the remainder in five equal yearly instalments. Mitchelstown Deer Park wood was sold in 1758 for ready money or a year's grace on good security. Ballyfoyle wood, near Kilkenny, went for sale in 1755 for half the purchase money down and the rest by agreement. Lackenavorna woods, outside Nenagh, were offered for sale in 1760 on terms of a year to pay with security and legal interest.

Once the timber was sold there remained the problem of felling, and here again the terms offered to buyers varied. If only part of a wood was to be felled, the trees to be cut were marked with oil and red lead. The time allowed for felling varied from one to seven years and it was customary for the timber merchant to install his workmen in huts in the wood and to provide grazing for cows. Sometimes the owner let land to the merchant for the accommodation of his workers, but in other cases no charge was made. Wooden articles were often made on the spot, especially in remote areas.

The practice of advertising woods for sale in newspapers began about 1730, though it was not until the 1740s that it became relatively common. After 1770 the number of such advertisements rapidly declined, probably because of the decreasing amount of timber, but also possibly because of the change in layout of Irish newspapers. During the early and middle part of the eighteenth century the four-page newspaper devoted nearly three pages to closely packed advertisements of every kind in small type, but by the end of the century newspapers carried fewer and they were more spaciously laid out. The advertisements were generally short and gave little information early in the century, but as time went on they became more detailed and often contained four relevant pieces of information: the species of tree, their age, the acreage, and distance from the nearest town. Less frequently a specific number of trees was mentioned and the size of timber they would yield.

Out of approximately 400 advertisements just under half give the acreage of wood for sale. The total area for each of the

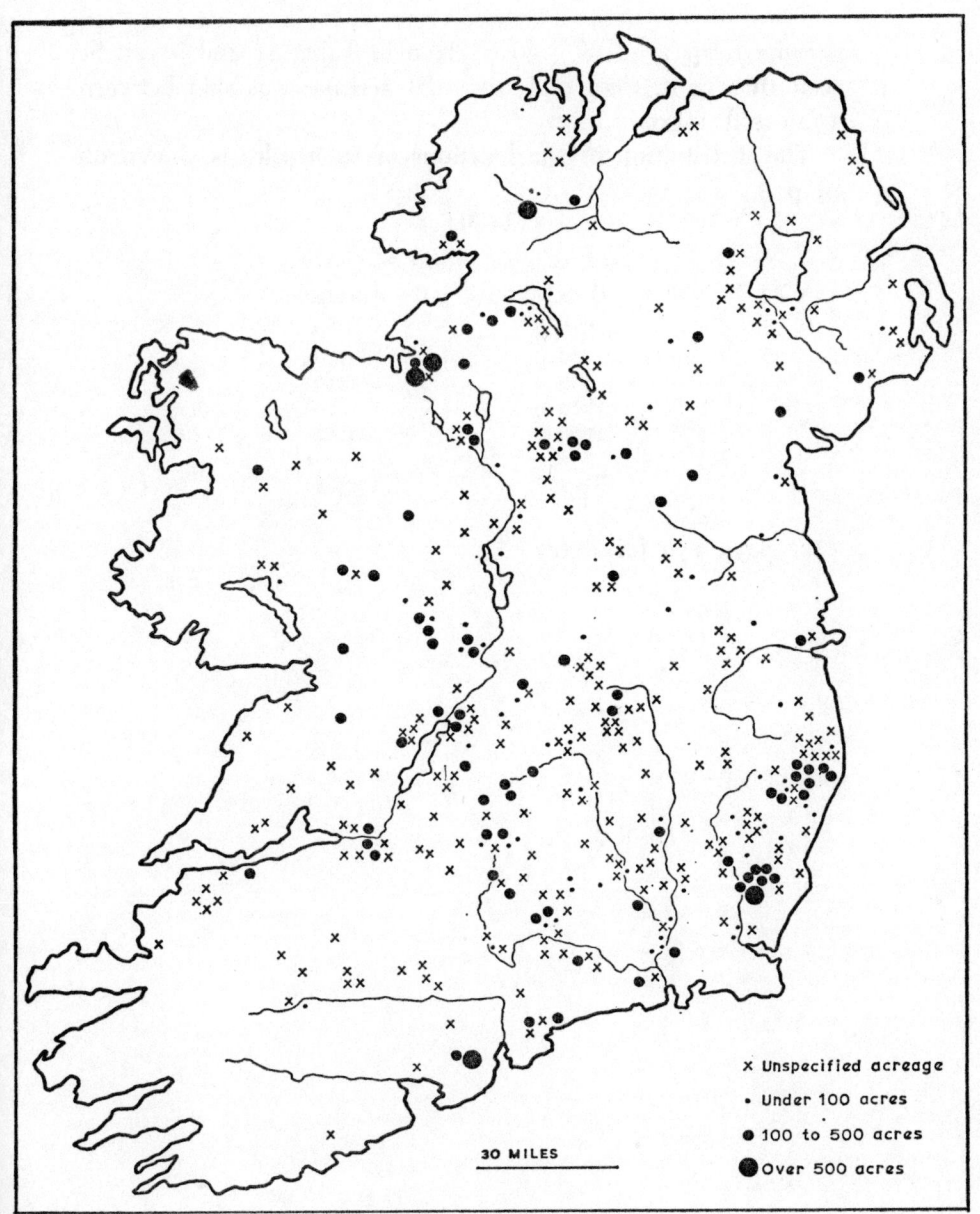

Map 7. Woods advertised for sale, 1730–80

decades from 1730 to 1780 is given in Table 21 and it can be seen that more than half the total acreage was sold between 1740 and 1760.

The distribution of the locations of such sales is shown on Map 7.

TABLE 21

ACREAGES OF WOOD FOR SALE

Decade	Acreage
1730–9	3,960
1740–9	7,570
1750–9	8,320
1760–9	4,829
1770–9	2,782
Total	27,461

(See page 159 for note)

CHAPTER 8

The Era of Private Planting

It is self-evident that the planting of trees will not be undertaken until the political situation in a country is reasonably stable, so that he who sows will reap. Furthermore, until it becomes clear that nature's bounty is either failing or inadequate, there is no incentive to plant a crop that cannot be harvested for 100 years. Even in England, where from Bosworth Field (1485) to Edgehill (1642) political stability had been more or less maintained, planting was a post-Restoration phenomenon. In Ireland political security for landlords came with the victories of Derry, Aughrim, and the Boyne at the close of the seventeenth century and lasted until Gladstone's land acts of the late nineteenth century. These eighteen decades span the era of private planting in Ireland.

The initial estate plantings, and indeed most plantings, made before the end of the seventeenth century were shelter belts, orchards, and avenues, and they tended to 'follow the flag'. That is, they were first made in the settled areas round Dublin and Cork. The number of species native to Ireland is limited, and the early plantings understandably consisted of recently introduced exotics: sycamore, beech, walnut, lime or horse chestnut

—the status symbol is not a twentieth-century innovation. Hand in hand with this initial ornamental planting went hedgerow planting. The division of the countryside into fields bounded by hawthorn hedges with regularly interspersed trees was an outward sign of the increasing anglicisation of Ireland. From the mid-seventeenth century one finds increasing numbers of leases binding a tenant not only to make a substantial hawthorn hedge round his holding but to plant and maintain trees in it. Nearly always ash was explicitly named and occasionally oak—both good utilitarian trees.

From newspaper advertisements it is clear that planting blocks of exotics began about 1700. To quote one result of this planting: there were twenty acres of forty-year-old fir for sale at Ireland's Grove near Portarlington in 1740.

The Dublin Society, formed in 1731, in its endless pursuit of good causes, found in planting a subject highly acceptable to its members, most of whom were landed gentry or related to them. From 1740 to 1808 the Society offered yearly awards to anyone who put in a sufficient number of saplings. The species it sponsored changed during the years but not the proviso that the plantation must be fenced and the fencing maintained for ten years. To start with the Society offered a flat £50 for the greatest number of trees planted, but by 1744 they were specifying oak, ash, elm, beech, walnut, and chestnut. During the 1760s silver and spruce fir, Scots and Weymouth pine, larch, Norway maple, sweet chestnut, and black cherries were added. During the 1780s the list was extended to include purple and copper beech; scarlet maple; American elm and birch; scarlet, turkey, prickly, Turner, Luccombe, American swamp, champagne, black, and white oak; cedar of Lebanon; the tulip tree; Athenian poplar; black larch; whole-leafed ash; cembo and swamp pine; Newfoundland and hemlock spruce; and two-thorned acacia.

Apart from gold and silver medals, the Society paid out £19,000 between 1766 and 1806 in premiums on 55,000,000 trees. As the trees were closely planted the acreages involved were small, an average of about ten per planter in any one year.

The Era of Private Planting

Considerable prestige was attached to winning a medal or a premium, though it was said that many who received such awards did not bother to maintain their fencing.

As well as voluntary there was some compulsory planting. Between 1698 and 1791 seventeen parliamentary acts were directed at conserving and increasing the area under woodland. The earlier acts tended to be repressive—it was forbidden to keep goats other than on mountain land, and it was illegal to make or sell articles made of oak or to make May poles. The penalty for cutting trees in daylight was the value of the timber plus a £2 fine, and if they were cut at night it was a felony. Planting was made compulsory for those holding leases of over thirty-one years, for ironmasters, and for holders of over 500 acres. Fines were to be inflicted for failure to plant and the enforcement of planting was put in the hands of grand juries. The 1721 act removed the penalties for failure to plant but increased the penalties for illegal felling. It also gave a tenant the right to a third of the trees he had planted during the term of his lease. Later acts further improved the position of a tenant until, in 1783, those with leases of over fourteen years could cut or sell any trees they had planted.

On the whole the grand juries took their supervisory duties seriously and did as they were instructed, allotting to each parish a number of trees to be put in each year. But the total number of trees was very small: the original act of 1698 aimed at 260,000 in thirty years spread over the whole of Ireland, which was just token planting. As has been suggested earlier perhaps Ireland was not so short of timber at the end of the seventeenth century as has been generally supposed. It may be that in spite of the gloomy preamble to the 1698 act, 'Forasmuch as by the late rebellion in this kingdom and the several iron works formerly here the timber is utterly destroyed, so that at the present there is not sufficient for the repairing of the houses destroyed, much less a prospect of building and improving in after times . . .', its purpose was primarily to protect landlords against tenants during a time of political unrest—750,000 acres changed hands as a result of the Glorious Revolution—and also to protect new owners against rapid cutting by the former landlords.

Eighteenth-century estate papers are full of complaints about tenants, and others, cutting timber. The agent on the Olivers' Limerick estate commented on the theft and mutilation of some ash trees, 'The sight of the same is scandalous to be seen'[1]. Some landlords felt strongly enough about the depredations made on their woods to offer rewards for the discovery of the offender: Philip Crosby offered 20 guineas—the yearly wage of four woodrangers, or Lord Gosford's butler—for information after some oaks were cut in his woods at Stradbally in Leix. The Duke of Leinster at Carton, Kildare, offered no reward for information in 1774 but simply announced that he had set man-traps in his plantations. The duke took his planting seriously—most estates did with a gardener and a woodranger but at Carton there was 'Jacob Smith, Planter to the Duke of Leinster'.

The tenant planting under a grand jury order, or an estate owner hoping to win a gold medal from the Dublin Society, were equally faced with the difficulty of finding young plants. The grand jury orders specifically forbade the planting of saplings from a growing wood, and anyway the ambitious landlord would not find many of the species he wished there. To begin with, young plants were imported from England but they frequently arrived at their destination in a sorry state, and so the eighteenth century saw an increasing number of nurseries established in Ireland. Many landlords had private nurseries and a few of them sold saplings to the public, but the need for supplies led to the opening of commercial nurseries. One of the earliest known belonged to William Griffith of Carlow town, whose stock in 1741 included Scots pine and silver and spruce fir. The Dublin Society encouraged nurserymen by giving premiums on the number of trees sold and paying fees for apprentices, and from 1740 to 1800 over ninety nurseries appeared in the provinces and sixty in or round Dublin[2].

The distribution of nurserymen and seedsmen in Dublin in

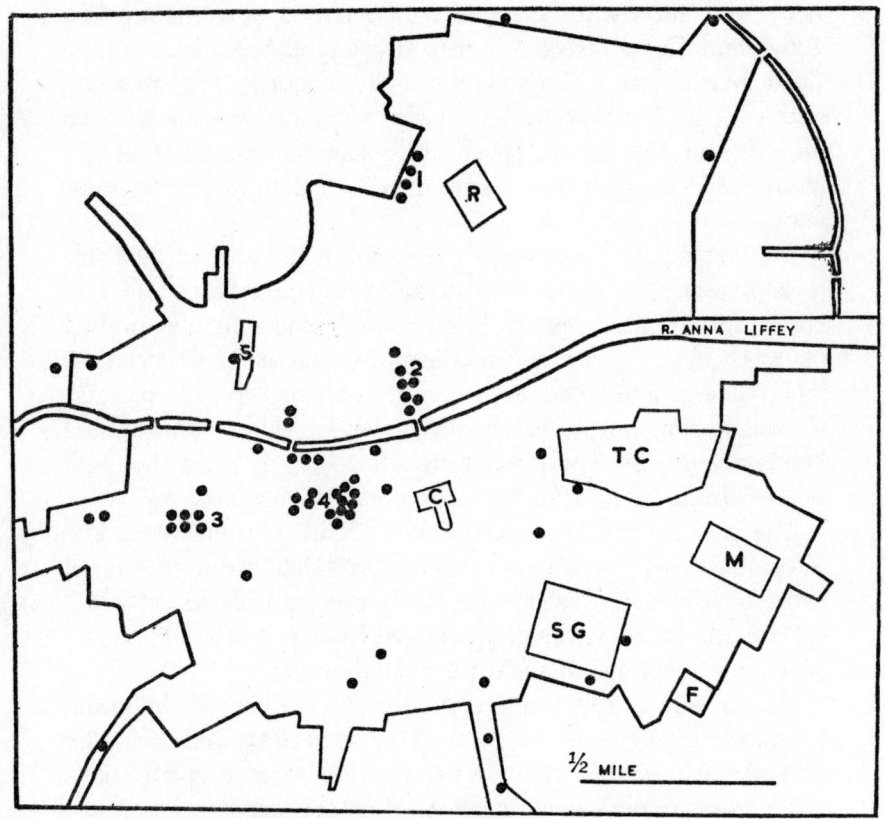

Map 8. Dublin seedsmen and nurserymen, 1740–1800

- R   Rutland Square
- TC Trinity College
- S   Smithfield
- C   Dublin Castle
- M   Merrion Square
- F   Fitzwilliam Square
- SG St Stephen's Green
- 1   Dorset Street
- 2   Capel Street
- 3   Thomas Street
- 4   Corn Market and Christchurch area

the second half of the eighteenth century is shown on Map 8. They were mostly to be found in the streets near the western quays and Capel street, Thomas street, and the streets round Corn Market and Christchurch. The majority, possibly all, sold young plants as well as seeds. Most of the nurseries on the edge of the city supplied the shops in the heart of the town, and often the shop and nursery belonged to the same man.

Contemporary nurserymen's lists show the species available to a planter. The hardwoods included weeping birch and ash, laburnum, spindle, rowan, English and American elm grafted on wych elm, American blackthorn, ash-leafed maple, arbutus, lime, sweet and scarlet chestnut, various species of oak, Canadian elm, purple beech, black poplar, balsam poplar, and Lombardy poplar. The prices ranged from 2s 6d for a thousand ash seedlings to 2s 8½d each for a scarlet chestnut.

The conifers available were fewer: Scots, cluster, stone, and Weymouth pine; spruce and silver fir; larch; balm of Gilead; cypress; yew; cedar; arbor vitae; and American black and white spruce; and for these the prices ranged from 4s 4d for a thousand Scots pine seedlings to 1s 6d for a single cedar.

As a result of the 100 years' peace, the efforts of the grand juries, the carrots held out by the Dublin Society, and the desire to embellish estates, there were 132,000 acres of plantations apart from natural wood in 1801. How this acreage was distributed according to species is shown in Table 22.

TABLE 22

ACREAGES OF PLANTATIONS[3]

	1801	1841
Ash	2,800	6,000
Beech	1,300	3,300
Elm	620	1,400
Oak	25,300	29,500
Conifers	2,800	25,000
Mixed	99,000	280,000
Total	131,820	345,200

The Era of Private Planting

By 1841 there were 345,000 acres of plantation, and a pointer to things to come, the area under conifers had increased eightfold. In the nineteenth century the conifer was planted in great numbers. Just as the orchid-hunters scoured the tropics for the Victorian hothouses, so the conifer-hunters searched the cordilleras on the west of the American continent for the estate arboretas. Four conifers were successful in Ireland—Douglas fir, Sitka spruce, Lodgepole pine *(Pinus contorta)*, and Noble fir.

From 1840 to 1880 cutting balanced planting and the acreage under plantations remained the same. The year 1880 marked the zenith of woodland acreage in private hands for the great land act of the following year began to transfer land control from landlord to tenant. The estate owners, conscious now of their insecurity, not only ceased to replenish their plantations but sold much of the existing timber to travelling sawmillers who came over from Great Britain and moved across the country from estate to estate like arboreal pests. World War I followed soon afterwards, and, as in England, so much timber was cut that by the 1920s there were in Ireland only about 130,000 acres of woodland, roughly a third of the acreage a century before. The process which had been accelerated at the end of the sixteenth century was completed and about a half of one per cent of the land remained as forest.

(See page 160 for notes)

CHAPTER 9

State Planting

Edlin has commented that if one looks at a map of the state plantations in Ireland it gives the impression of having been dusted with a pepper pot[1]. This comment is true in that the planted areas are small and scattered: the largest single area —Cloosh valley in Galway—was 10,800 acres in 1963. Most of the forests are between 2,000 and 3,000 acres and small by English or continental standards. And it is discerning in so far as it suggests a random distribution. The acquisition of land was rigidly determined by economics. Until 1969 £10 an acre was the ceiling price. As one land acquisition officer said, 'If I go to buy land and see a couple of sheep on it I know I'm wasting my time because the land's worth more to the farmer than the government is willing to pay'. So the land available for forestry —apart from a few exceptions where old demesnes have come into the hands of foresters—has been on the whole poor marginal or upland areas and, since 1950, because of improved planting techniques, bog land.

Although state planting in Ireland goes back to 1903 it might have begun twenty years earlier. In 1883, two years after his 1881 land act, Gladstone set up a commission to consider the possibilities of reafforestation in Ireland. Daniel Howitz[2], a Danish forester, spent three months in the country and concluded that 5 million acres, or a quarter of the land, was more fitted for forestry than anything else, and that of these 5 million acres 3 million should be planted on the poor agricultural land along the western seaboard, where the forests could provide work for the impoverished rural population, and a million acres in the main catchment basins to control soil erosion and runoff. He envisaged this programme as spread over half a century, and advised planting very many species initially to discover which were best suited to local conditions[3]. Howitz's opinions were more or less in line with those expressed by Sir Arthur Griffith, who in his Valuation Report of 1845 suggested that half of the 6 million acres of waste and unproductive land could be planted.

The first two acquisitions made by the state were curiously enough of considerable interest in the history of forestry[4]. They were Avondale in Wicklow, the home of Samuel Hayes, who wrote the first Irish book on planting in 1794 and who helped to choose the site for Dublin's Botanical Garden; and Walworth wood, near Ballykelly in Londonderry, once owned by Sir William Walworth, the Lord Mayor of London who struck down Wat Tyler in 1381, which had a long history of planting behind it.

Between 1906 and the early 1930s forestry, first in the whole of Ireland and then in the Irish Free State, was under the care of Arthur Charles Forbes (1865–1950), who had been forester in charge of the Longleat estate belonging to the Marquess of Bath and later lecturer in Forestry at the Armstrong College of Science, Newcastle. He was appointed editor of the newly established *Quarterly Journal of Forestry* in 1906 and his name appears as editor on the first number[5], but he did not in fact act as editor for in the same year he was brought over to Ireland. During the quarter of a century he directed forestry operations in Ireland, 26,900 acres were planted.

Forbes was succeeded in 1934 by Dr Mark Loudon Anderson

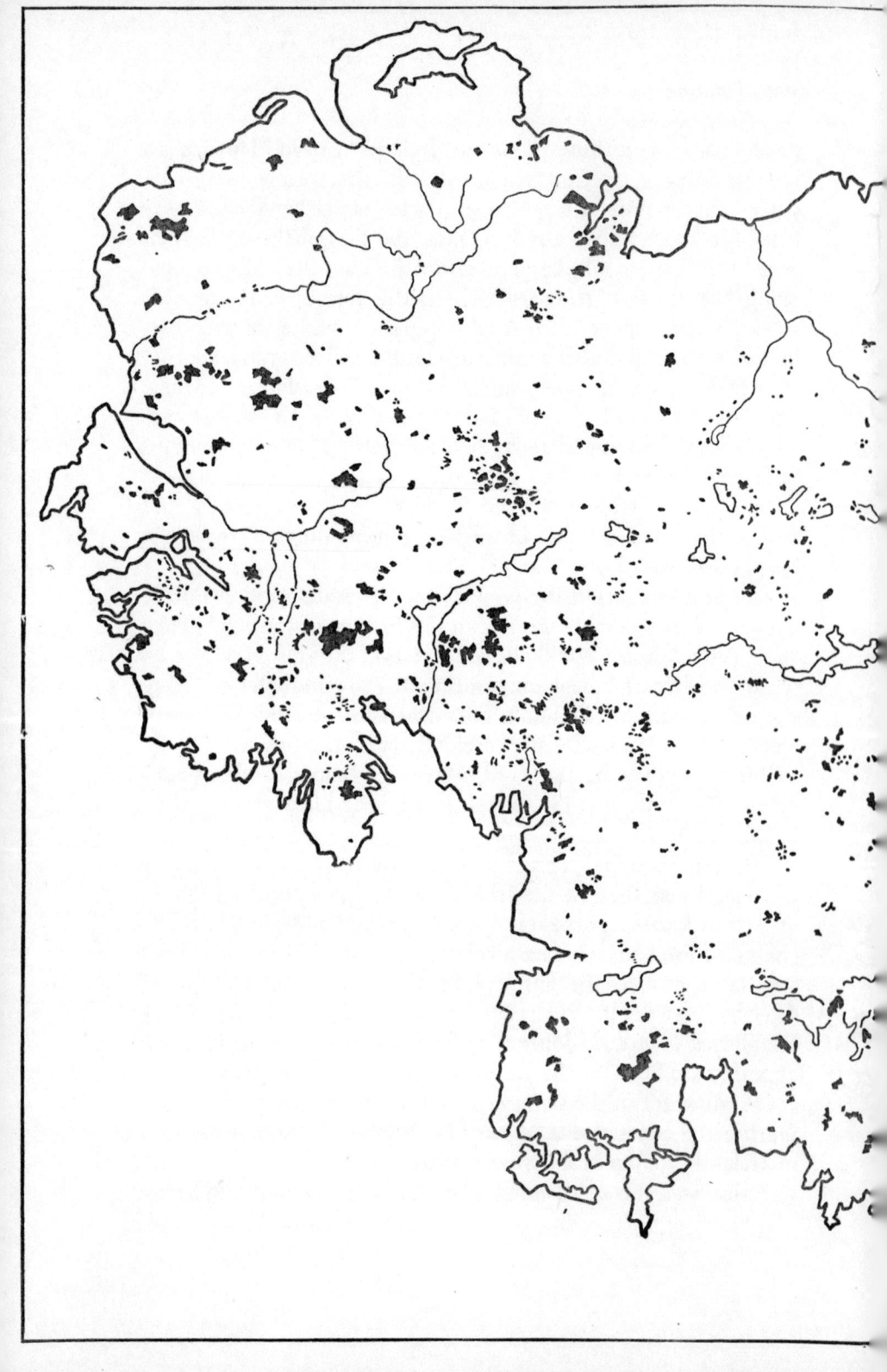

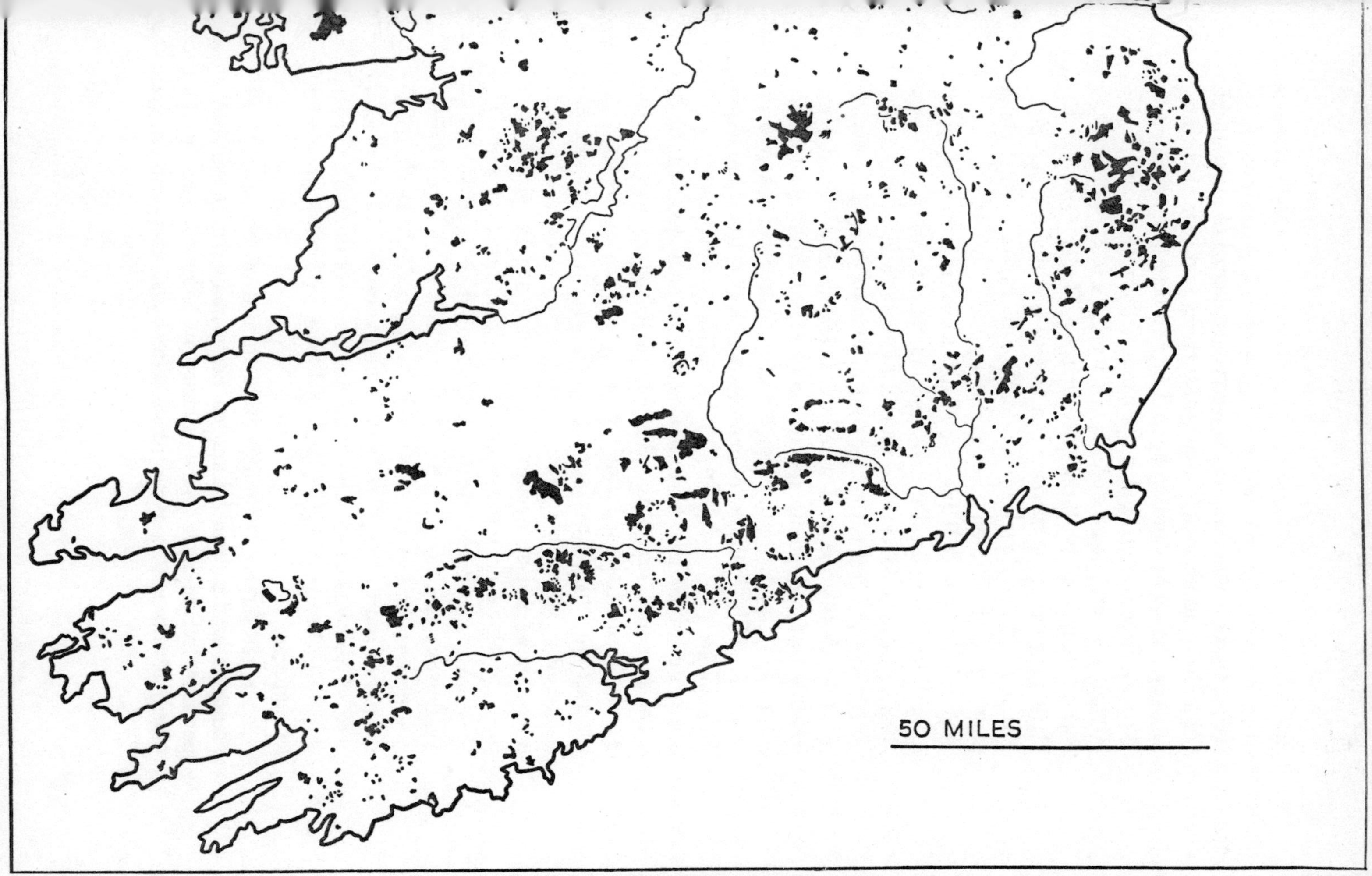

Map 9. State forests, 1966

50 MILES

(1895–1961), who was a forestry graduate of Edinburgh University and ultimately returned there as Professor of Forestry in 1951. Before he came to Ireland he had been with the British Forestry Commission and a Research Officer in Scotland. He published several books and his last work, A *history of Scottish Forestry*, appeared posthumously in 1969. Under Anderson the planting programme increased in Eire to just under 8,000 acres a year by the outbreak of World War II. In Northern Ireland planting reached 1,000 acres a year by the same date.

Since World War II the rate of planting has been stepped up in both the Republic of Ireland and Northern Ireland: it is now about 25,000 acres a year in the Republic and 5,000 in the North. The result was that by 1968 after sixty-five years of state planting there were half a million acres of land acquired for forestry in the Republic and 91,000 acres in Northern Ireland, rather less than twice the area private planters had achieved by 1840. Roughly half the present forests are of trees under ten years old.

Private woodlands have not completely disappeared; they cover 60,000 acres in the Republic and 30,000 acres in the North. Most of the old estate woods are of deciduous trees, in contrast to the state forests which are almost completely coniferous; these old woods are often untended, wild, uneconomic, and an eyesore to some, but not all, professional foresters. However, some people like them: they look 'natural', they give a softness to the landscape, they bear the same relationship to a properly tended forest as a patch of wild wayside flowers does to a geometrical plot of bedded-out plants. Is Ireland so short of land, so poor, that she cannot afford 90,000 acres of old woodland? Less than 90,000 acres in fact, for part of it is modern planting, encouraged by a state grant of around £20 an acre.

As regards species, conifers have formed 96 per cent of all plantings in the Republic and 94 per cent in the North. Both parts of Ireland have given priority to Sitka spruce: in the North it occupies 55 per cent of state forest areas and in the

State Planting

Republic a little over 40 per cent. After agreeing on Sitka spruce as a desirable tree to cover approximately 250,000 acres of Ireland, the two countries part company; the Republic's second most widely planted tree is *Pinus contorta* or Lodgepole pine, so called because some of the North American Indians used it for the poles of their teepees. This covers just under a third of forests in the Republic. Considerable difficulty was found in locating a suitable strain of *Pinus contorta* for seed. One particularly unsatisfactory source of origin was Lulu Island—so named after the favourite performer of a Vancouver Island governor—and better results were obtained with seed from the Washington Coast region and the Olympic peninsula. In the North second choice is Norway spruce, which is used in about 10 per cent of plantations, and it occupies the same percentage of forest land in the Republic. Other conifers which are used to a limited extent are Monterey, Scots and Corsican pine; Japanese, European and hybrid larch; Noble, Giant and Douglas fir; Western red cedar; and Western hemlock.

Beech comes first among the hardwoods in the Republic, followed by birch. In the North beech and oak occupy about the same space. Small quantities of ash and sycamore are also planted.

It is difficult to determine from official figures how far homegrown timber meets local needs. The annual consumption in the Republic is about 15 million cubic ft (1965), of which about half is Irish wood. Sawmills appear to take about 3 million and processing factories about 4 million cubic ft of Irish timber. There is also a small export trade from the Republic valued at £1,500,000 yearly, mostly of processed wood and veneer sheets. The value of imported wood in 1965 was £9,500,000, of which £1,000,000 was for hardwoods.

In the North the output is of course considerably smaller, about 1¾ million cubic ft in 1963.

Until recently if one had asked 'What is the raison d'etre of these state forests?' the answers would have been clear cut: they used up land unsuitable for other types of agriculture, they provided employment in 1967-8 for about 4,000 men in the Republic and 2,000 in the North of Ireland, and some would perhaps have added that it was 'a good thing' to plant trees. But today there is a change of opinion and there are those who do not consider either the siting of the forests or the devotion of 95 per cent of all planted area to conifers 'a good thing'.

Take the conifers that have been brought in. Sitka spruce is native to a narrow coastal strip about fifty miles wide which runs from the east end of Kadiak Island in Alaska to north California. Here, in its natural habitat, it can grow to 280ft and live for 800 years. Douglas fir, less widely planted, also comes from the American cordilleras and there can reach over 400ft and live for 1,000 years. Norway spruce belongs to northern Europe and Asia and also high mountain ranges in those continents. These exotics, deprived of the long resting period they normally experience in their native high altitudes or high latitudes, grow extremely quickly in the damp, relatively warm Irish climate and are cut at about forty years. These trees are forced like battery hens and their timber is comparable in quality to the eggs and flesh derived from those hens. Battery eggs are a poor parody of eggs from the farmyard hen, but they are said to be a more economic proposition to the producer and as such are thrust, willy-nilly, on the consumer. You can have more poor-flavoured eggs or fewer well-flavoured eggs for the same money—man cannot ultimately defeat nature's balance.

State forests were until recently graded quality class I to class V. The innocent may well believe that this referred to the quality of the timber, but not at all; it referred to the volume of timber per acre, quality class I being the highest grade in that the trees on such a stand put on the maximum volume in a given period. It is a long time since the fattest pig was considered the best bacon producer.

It is significant that Scots pine, a conifer native to the British Isles, does not thrive in the wetter regions in Ireland, and the upland regions are the wetter regions. Its rate of growth, 8 per cent in the first twenty years as against 30 per cent in the exotic conifers, highlights the abnormal conditions under which the latter are grown.

Official figures do not provide the cost per cubic ft of home-grown softwood, but the very climatic features which enable them to grow with such rapidity make extraction costs high, as does the siting on hilly land, usually far from their ultimate destination, make transport costs high. One is entitled to ask 'Is the production of softwood an economic proposition or is it in fact a state-subsidised industry?' In the Republic in 1967–8 expenditure amounted to £4,250,000 and income to £707,000. The corresponding figures for Northern Ireland were £1,400,000 and £375,000. If it is being subsidised, to what end is it being subsidised?

As well as the economic side to state forestry there is the question of amenity. The motor car has come to be used as a mobile sitting room—as one sits in the sitting room at home and looks at television so one sits in the car and looks at the countryside. People want to look at what pleases them. If afforestation continues at its present rate of 30,000 acres a year, does this mean that in time the hills will be one vast coniferous plantation, and if so is that what the taxpayer wants? Are the mountain roads from which at present one can see miles over the open hills going to turn into long dark tunnels bounded by parallel lines of Sitka spruce or *Pinus contorta*? Will the flat boglands with their changing colours be sheets of dark green that know no season?

In the past twenty years there have been moves made to open certain forests to the public. Originally amenity meant little more than putting a row of hardwoods along the side of a coniferous plantation, it being recognised that people were not charmed by the raw edge of an unthinned block of conifers; but amenity now means a great deal more, and to date has taken three forms. Forest scenic drives have been laid out, as at Slieve Gullion in county Armagh and Glendalough in Wicklow. The 1953 Forestry Act of Northern Ireland provided for the establishment of National Forest Parks and the first of these was made at Tollymore Forest Park in county Down. Tollymore is an old estate with an arboretum and 130 acres of old woods. The number of cars entering the Park increased from 7,000 annually in 1955 to 33,894 in 1964, most of the visitors living within a radius of forty miles[6]. Gortin Glen in Tyrone has been opened since and more are to follow. Legislation in the Republic does not provide specifically for public parks, but some of the forests are open to the public, and the John F. Kennedy Memorial Forest Park in Wexford is being laid out as an arboretum.

The third type of amenity provided by the forestry services may loosely be called educational. It includes labelling trees with their names, laying of nature trails for children, and providing hides and look-outs for deer- and badger-watchers, as has been done at Portumna. Wild life Officers have also been appointed.

Admission of the public can make difficulties for those in charge of a forest: people leave litter, and, in spite of appeals, do light fires and drop lighted matches and cigarette ends, and children are destructive. The danger of 'commercialisation' has also to be guarded against. To take a case in point: Gosford Castle Forest in Armagh is open to the public and a large car park, a café, and toilet facilities have been provided; but attached to the estate is what was once a magnificent walled garden now derelict and pressure is being put on forestry officials to turn it into a children's playground with swings, roundabouts etc, 'because there is nothing for them to do in the place'. It could be made into a maze, and the use of mechanical hedge shears should not make its upkeep unduly costly.

Desirable as scenic drives and forest parks are both as amenities and as honey-pots to concentrate visitors in controlled areas they are not enough. The time has come for the framers of forest policy to consider their responsibilities. When a few thousand acres a year were planted it did not matter very greatly where they were located, but at the present rate of planting in our children's children's time there will be very little open moors left on the hillsides. Howitz's opinion that a quarter of

Ireland could be forested is now a possibility, thanks to the use of heavy ploughs powered by crawler tractors that have in places pushed planting up above 2,000ft, the use of fertilisers, and the elimination of competing vegetation by chemicals.

It is true to say that the future landscape of Ireland lies in the hands of the forest planners. Have they a plan or is land being acquired and planted piecemeal, regardless of its site? The Irish people are directly concerned for Ireland is so well provided with roads that almost literally nowhere is out of sight of a road. Once they leave the lowland, do they wish to pass into a uniform area of coniferous forest that blocks all vistas? Planning the hill landscape of Ireland does not just mean avoiding rectangular patches of one species, seeing that the skyline is not serrated like the edge of a saw, and camouflaging the ugly streak that a straight forest road can make through a stand. It does not just mean providing the odd forest park or scenic drive that one has to go specially to see. It means considering the whole sweep of country over several square miles and deciding where a forest can best be placed so that it adds to the landscape. It means reducing the 95 per cent of conifers now forming the planted areas and using more hardwoods, so that there are areas of Ireland's own native trees even if more has to be paid for the richer soil that hardwoods need. So few deciduous trees are being planted that, unless there is a deliberate drive, not only will miles of hill land be swathed in conifers and the scenery altered but the face of the lowlands will also be changed.

The lowland scenery as we know it was created during the past two or three centuries by the planting of trees for ornament, shelter belts, and in hedgerows; and though statistically the lowland area under wood is small the general effect in many parts is of a comfortably wooded countryside. There were 18 million hedgerow and ornamental trees in 1841. But trees are mortal—even the oak, contrary to popular belief, has a life of only two or three centuries—and as these old trees fall or are cut they are replaced, if replaced at all, by conifers that will be quite big in twenty years. The old estates with their lime, sycamore, cedar, beech, redwoods, wellingtonias, and monkey puzzles are part of our heritage and should be preserved, for we shall not see their like again. There is no reason why a state grant should not be made if they are still in private hands to ensure their proper upkeep, and if they pass into government hands, as some do, there is no need to clear fell them and put a conifer plantation in their place. It is not considered unreasonable to spend public money on the upkeep of ancient buildings nor would it be unreasonable to spend public money on the upkeep of these old collections.

Until a few years ago the hedgerow trees seemed to be more secure. The hedger was more often than not a countryman who moved slowly along the hedge and spared the odd sapling so that the hedgerow retained its timber; but now he arrives from the nearest town in his car, sits in a covered tractor and sweeps along the line of the hedge with a mechanical hedge-cutter.

Hardwoods are economically unprofitable: they demand better soil than conifers, they will not grow in the exposed positions conifers will tolerate, there is at the moment only a small demand for their timber, and they take a century to mature. Nevertheless, if rural Ireland is to retain its distinctive landscape hardwoods will have to be included in any forestry planning for the country.

(See page 160 for notes)

Notes and References

ABBREVIATIONS

A. *Libraries etc*
    BM   British Museum
    NLI   National Library of Ireland
    PRO   Public Record Office, London
    PROI   Public Record Office, Ireland
    PRONI   Public Record Office, Northern Ireland

B. *Journals etc*

Ardagh CASJ	Journal of the Ardagh and Clonmacnoise Antiquarian Society
Carew MSS	Calendar of the Carew manuscripts
CSPD	Calendar of the State Papers Domestic
CSPI	Calendar of the State Papers Ireland
Cork HASJ	Journal of the Cork Historical and Archaeological Society
Galway AHSJ	Journal of the Galway Archaeological and Historical Society
Kenmare MSS	The Kenmare manuscripts, ed MacLysaght
Kil ASJ	Journal of the Kilkenny Archaeological Society
Louth JA	Journal of the Louth Archaeological Society

OSMSS	Ordnance Survey manuscripts
OS letters	Ordnance Survey letters
Phillips MSS	Londonderry and the London Companies, ed Chart
PRDS	Proceedings of the Royal Dublin Society
PRIA	Proceedings of the Royal Irish Academy
RHASIJ	Journal of the Royal Historical and Archaeological Society of Ireland.
RSAIJ	Journal of the Royal Society of Antiquarians of Ireland
Shapland-Carew MSS	Shapland-Carew papers, ed Longfield
UJA	Ulster Journal of Archaeology
Waterford ASJ	Journal of the Waterford and South East of Ireland Archaeological Society

Chapter 1 (pages 15-32)

1 *House of Commons' Journals*, 47, (1792), 327
2 MacLysaght, *Irish life in the seventeenth century*, 396
3 Lucas. 'Sacred trees of Ireland', *Cork HASJ*, lxviii, 32 et seq
4 *Carew MSS*, 1589–1600, 20; *CSPI*, 1615–25, 512
5 'Extracts from the Journal of Thomas Dineley', *Kil ASJ*, 1, pt 1, New Ser, 176
6 Hill. *Stewarts of Ballintoy*, 51–2
7 Petty. *Political anatomy of Ireland*; Threlkeld. *Synopsis Stirpium Hibernicanum*
8 Mitchell. 'Littleton Bog, Tipperary', *Geol Soc America* (1965), 1–6
9 Lucas, op cit
10 *An Claidheanh Soluis*, x no 10 (1908), 7
11 Joyce. *Irish local names explained*
12 Murphy. *Early Irish lyrics*, 11
13 Meyer. *Selections from ancient Irish poetry*, 47
14 Harris. *Hibernica*, 50
15 Hore. 'Woods and fastnesses in ancient Ireland', *UJA*, 1st Ser, vi, 154
16 *CSPI*, 1601–3, 259
17 Bagwell. *Ireland under the Stuarts*, i, 82–3
18 Marshall. *History of Charlemont Fort*; 42
19 Moody. 'Redmond O'Hanlon', *Proc Belfast Nat Hist and Phil Soc*, 2nd Ser, 1, pt 1, 17–23

20 Dunlop. *Ireland under the Commonwealth*, ii, 178, 192, 481, 664
21 Prendergast. 'The Plantation of Idrone', *Kil ASJ*, New Ser, iii, pt 1, 76
22 Westropp. 'Ring forts in Moyarta barony', *RSAIJ*, xxxix, pt 2, 156
23 PRONI, Brownlow papers, T 951, 971
24 *Concise view of the Irish Society*, 115
25 PRONI, Abercorn papers, Cal 95 A

Chapter 2 (pages 35-36)

1 McCracken. 'The woodlands of Ulster in the early seventeenth century', UJA, x (1947), 15-25
   'The woodlands of Ireland c. 1600', *Irish Historical Studies*, xi no 44 (1959), 271-96
   'Irish woodlands, 1600 to 1800', *Quarterly Journal of Forestry*, lvii no 2 (1963), 95-105
2 Atkinson. *Ireland exhibited to England*, ii, 205; OSMSS, Ardclinis parish, box 2
3 CSPI, 1606-8, 211; 1608-10, 16
4 *Carew MSS, 1601-3*, 315
5 *Memoirs Geological Survey to sheets 4, 5, 9-11, 15-16*, 7
6 Armagh Public Library. Henry. 'Killasher Parish', 115-16
7 Wood-Martin. *Sligo*, 3-4.
8 OS letters, Galway, i, 196
9 Dutton. *Galway*, 25; Simington. *Books of Survey and Distribution*, i, Roscommon
10 Irish Folklore Commission, ccv; 202
11 Wakefield. *Ireland*, i, 535
12 Hore, op cit, 157
13 Smith. *Cork*, 201
14 Hodson. 'Woodlands of West Cork', *Cork HASJ*, Ser 2, viii, 115
15 Otway. *Sketches in Ireland*, 322
16 Simington. *The Civil Survey*, vi, Waterford
17 *Cork HASJ*, Ser 2, xxi, 199
18 CSPI, 1608-10, 88
19 CSPD, 1671, 207, 529
20 Le Fanu. 'The royal forest of Glencree', *RSAIJ*, xiii, pt 3, 268, 274
21 Prendergast. 'The tory war of Ulster', *Kil ASJ*, new ser, vi, 35

22 Carew MSS, 1603-24, 381
23 Hodson, op cit
24 O'Donovan. *Antiquities of Londonderry*, 93
25 OSMSS, Londonderry, parishes of Balteagh, Banagher and Boveagh
26 Otway, op cit, 335

Chapter 3 (pages 57-96)

1 Longfield, *Anglo-Irish trade*, 211

Coöpering

1 2 Geo I, c 16; 10 Geo I, c 29; 12 Geo I, c 25
2 *House of Commons' Journals*, vol 48, 292, 297, 312-13
3 *Minutes of the Dublin Society*, xii (1775-6), 193
4 O'Sullivan. *Economic history of Cork*, 121, 170; MacLysaght, op cit, 398-9

Shipbuilding

1 Lucas. 'The dugout canoe in Ireland. The literary evidence', *Sartruck ur Varberg museum* (1963)
2 Dowdall. 'A description of Longford', *Ardagh CASJ*, 1 no 3, 25-6
3 *Faulkner's Dublin Journal* (1753)
4 CSPI, 1608-10, 21, 88, 225; 1611-14, 132, 369
5 Carew MSS, 1603-24, 194
6 CSPI, 1666-9, 147, 417, 727
7 Carter, B., Letter to *The Times*, 3 August 1963
8 Nelson's *Victory* built in 1765 for £53,000 only survived because she was extensively repaired. By 1800 the navy had spent £189,000 on her, exclusive of the original cost
9 Fincham. *History of naval architecture*, 76-7, 211-12
10 CSPI, 1660-2, 166; 1663-6, 524, 612
11 CSPI, 1666-9, 610
12 CSPI, 1611-14, 369; 1615-25, 269; 1633-47, 85; *Danver. Letters . . . East India Co.*, i; Rpt select committee on industries (Ire), 745; Smith. *Cork*, 121, 201
13 PRONI, T 615; CSPI, 1660-2, 429, 1666-9, 666
14 NLI, MSS 8110; Benn. *Belfast*, 309-11
15 Simms. *The Jacobite parliament*, 17; O'Brien. *Ireland in the seventeenth century*, 171
16 Benn, op cit, 354-6

17	Wilson's *Dublin directories* for various years; *Freeman's Journal*, Feb 1764, May 1765, Aug 1765, July 1771; *Faulkner's Dublin Journal*, July 1766, Mar 1767, Apr 1768
18	*Faulkner's Dublin Journal*, Feb 1778
19	Shapland-Carew MSS, 92, 99, 100, 123, 126, 128
20	Cork directories for various years
21	*Faulkner's Dublin Journal*, Aug 1768; *Limerick Chronicle*, Jan 1771
22	Tighe. *Kilkenny*, 141; Weld. *Killarney*
23	Wakefield, op cit, ii, 63
24	Customs returns
25	CSPI, 1666–9, 584–5; Benn, op cit, 310

Housebuilding

1. *Dublin Penny Journal*, 15 Sept 1832
2. OSMSS, Londonderry, Ballyscullion parish
3. Carew MSS, 1603–25, 60; Moody. *Londonderry Plantation*, 136
4. Henry. 'The woods and trees of Ireland', *Louth ASJ*, iii no 3, 243; CSPI, 1647–60, 195; *A concise view of the Irish Society*, 23–4; Phillips MSS, 67, 69
5. Hill. *The Plantation of Ulster*, 57
6. Montgomery MSS, 302; Young. *Old Belfast*, 41; Atkinson. *Donaghcloney parish*, 31, 87, 89; CSPI, 1666–9, 156
7. CSPD, 1667, 531
8. CSPI, 1669–70, 248
9. Ibid, 140, 285
10. Dunlop, op cit, 523
11. 'Letter from Portadown', *UJA*, 2nd ser, iv
12. Graham. *Social life in Scotland*, 196–7; 4 Anne, c 9; Coote. *Armagh*, 134; PROI, Sarsfield-Vesey corr, letters 114, 194
13. Porritt. *The unreformed House of Commons*

Tanning

1. Atkinson. *Ireland exhibited to England*, ii, 273
2. CSPD, 1671, 184; *House of Commons' Journals*, vol 47, 276; Laslett. *Timber and the timber trade*, 115; Dutton. *Galway*, 451; Kenmare MSS, 419; Coote. *Kings county*, 59; Dutton. *Clare*, 280; Wallace. *Manufactures of Ireland*, 285, 293; NLI, MS 10934.
3. CSPI, 1669–70, 54.
4. *The Querist*, 48
5. CSPD, 1671, 184

6   Phillips MSS, 59; CSPI, 1625–32, 331; 11 Eliz, c 2; 10 Ch I, c 23; 4 Anne, c 9; 7 Geo III, c 23.
7   PRONI, T 473, 36, T 971, 39; Benn, op cit, 249, 283
8   Most of these sites are taken from newspaper advertisements
9   Wakefield, op cit, 721
10 Ainsworth reports, ix, 2189

Glassmaking

1   Hennessy. Raleigh in Ireland, 75–6
2   Westropp. Irish glass, 21, 24–6, 29, 31–2
3   CSPI, 1633–47, 37, 318
4   CSPI, 1647–60, 250; 1663–5, 602–3, 698
5   CSPI, 1669–70, 301–2

Ironmaking

1   Andrews. 'Notes on . . . the Irish iron industry', Irish Geography, iii no 3; McCracken. 'Charcoal burning ironworks in . . . Ireland', UJA, xx (1957), 'Supplementary list . . . ironworks', ibid, xxviii (1965)
2   McParlan. Leitrim, 72
3   Hayes. Practical planting, 112; Considerations concerning balance of trade between English and foreign iron, 2
4   McCracken, op cit, 123–5

Chapter 4 (pages 97–111)

1   A concise view of the Irish Society, 19
2   Carew MSS, 1575–88, 396
3   CSPI, 1666–9, 337
4   CSPI, 1611–14, 121
5   Stebbing. Raleigh, 161; Carew MSS, 1601–3, 109
6   CSPI, 1611–14, 1, 65
7   CSPI, 1615–25, 144; Caulfield. Council Book of Youghal, 80–2; Lismore papers, 1st ser, i, 185
8   Lismore papers; Court minutes . . . East India Co., 1635–9, 167, 1640–3, 69, 233; CSPI, 1615–25, 91
9   CSPI, 1608–10, 88
10 Brereton, Travels in . . . Ireland, 151
11 McCracken, 'The Irish timber trade in the seventeenth century', Irish Forestry, xxi no 1, 5
12 CSPI, 1633–47, 125, 250
13 CSPI, 1660–2, 429; Hayes, op cit, 111

14 Moody, op cit, 105, 113–14, 143, 146–9, 348, 362; CSPI, 1611–14, 300–1; *Phillips MSS*, 21, 27, 82, 89, 96
15 PRONI, T 522/29, TSP Ire, 6
16 McGrath. *Merchants . . . Bristol*, 284, 286–7, 289
17 CSPI, 1663–5, 693, 1669–70, 54; Harris. *Remarks on . . . trade of Ireland*, 22; BM, Add MSS 4759
18 10 Wm, c 11

Chapter 5 (pages 112–21)

1 The statistics in this chapter are compiled from Customs returns in the NLI, PRO, and the Library of the Customs and Excise, London.
   Contemporary accounts of Irish trade in the eighteenth century include Dobbs. *Essay on the trade of Ireland* (1729); Wallace. *Essay on the manufactures of Ireland* (1785); Sheffield. *Observations on the manufactures of Ireland* (1785); Newenham. *A view of the commercial circumstances of Ireland* (1809); and *The Dublin Society's Weekly Observations* (1756)
2 Dobbs, op cit, 364
3 Faulkner's *Dublin Journal*, Feb 1761, Aug 1769, Feb 1773, May and Sept 1774
4 Chambers's *Cyclopaedia*, 1738

Chapter 6 (pages 122–6)

1 Hayes, op cit, 114
2 CSPD, 1670, 583–7; 1671, 76–7, 135–6, 183–4, 207
3 CSPI, 1663–5, 649, 655
4 Customs returns
5 Jones and Simmons. *The story of the saw*, 23–4
6 CSPI, 1666–9, 358
7 Faulkner's *Dublin Journal*, Oct 1776
8 *Kenmare MSS*, 272, 279, 284; *Carew-Shapland MSS*, 115, 146; Sampson. *Londonderry*, 444; Ainsworth rpts, v, Headford papers

Chapter 7 (pages 127–34)

1 Wilson's *Dublin Directories*, 1752–1800. The information on timber merchants has been derived almost entirely from their advertisements in newspapers.

Chapter 8 (pages 135-41)

1. NLI, MS 10934, Oliver estate and family papers, 12 Dec 1775
2. McCracken. 'Irish nurserymen and seedsmen, 1740 to 1800', *Quarterly Journal of Forestry*, lix no 2; 'Notes on eighteenth century nurserymen', *Irish Forestry*, xxiv no 1
3. *Parliamentary Gazetteer of Ireland*, 1, lxxiii

Chapter 9 (pages 142-52)

1. Edlin. 'Review of the forests of Ireland', *Quarterly Journal of Forestry*, lxi no 2, 176
2. Howitz (1841-93) was the son of a land agent and graduated in forestry in Denmark in 1865. He became Forest Conservator and Superintendent for Victoria (Australia) and Danish vice-consul in Melbourne. In 1882 he advised the French government on forestry questions in Algeria. From 1887 to 1893 he was in Denmark doing anthropological work. He died in New York.
3. McNamara. 'Extracts . . . Rpt. by Howitz', *Irish Forestry*, xxiv no 2, 77-86
4. For an account of state forestry in Ireland see *The forests of Ireland*, 1966, edited by H. M. Fitzpatrick for the Society of Irish Foresters, which has also produced a journal, *Irish Forestry* biannually since 1943
5. Wright. 'Our Journal', *Quarterly Journal of Forestry*, lx, pt 1, 7
6. Kilpatrick. 'Public response to forest recreation in Northern Ireland', *Irish Forestry*, xxii no 1 (1965)

APPENDIX 1

Wood acreages in the mid-seventeenth century

These statistics of woodland and scrub have been compiled from the eleven published volumes of the *Civil Survey* and from the unpublished volumes of the Books of Survey and Distribution in the Public Record Office, Dublin. The figures give a partial picture only of the amount of woodland in Ireland at the time, as a complete survey was not made, and also in some cases, and this is particularly true of Roscommon, only a verbal description was set down without any information as to the area of woodland.

In the Surveys various terms—scrub, scrubby wood, woody pasture, pasturable wood, boggy wood, dwarf wood, underwood, rocky wood—are used to describe wooded areas which did not carry high forest. These terms have been grouped together under the blanket term of underwood.

The original statistics are in Irish acres and these have been converted into English acres.

A full description of the circumstances under which the Surveys were carried out is given by J. G. Simms, in 'The Civil Survey, 1654–6', *Irish Historical Studies*, ix, no 35.

County	Wood	Underwood	Total acreage
Carlow	1,800	490	2,290
Clare	11,580	33,350	44,930
Cork	7,690	6,860	14,550
Donegal		3,650	3,650
Dublin	90	650	740
Galway	4,900	9,620	14,520
Kerry	3,290	2,020	5,310
Kildare	980	5,190	6,170
Kilkenny	910	120	1,030
Leix	300	1,000	1,300
Limerick	5,890	11,390	17,280
Longford	380		380
Mayo	1,340	1,760	3,100
Meath	430	2,260	2,690
Offaly	500		500
Sligo	430		430
Tipperary	7,440	11,310	18,750
Waterford	4,810	8,920	13,730
Westmeath	2,730	330	3,060
Wexford	16,860	1,510	18,370
TOTAL	72,350	100,430	172,780

APPENDIX 2

Ships employed in the trade of Ireland, 1753

Port	Number	Tonnage			
		Irish	British	Others	Total
Baltimore	6	110	212		322
Belfast	468	5,425			16,263
Coleraine	26	630	210		840
Cork	732	8,781			50,515
Dingle	8	170	86		256
Donaghadee	12	75	260		335
Drogheda	200	1,036			7,276
Dublin	2,360	15,042			160,113
Dundalk	179	2,987			7,128
Galway	32	975			2,150
Killybegs	7	300			330
Kinsale	41	110			2,006
Limerick	80	1,310			6,060
Derry	100	6,580			8,175
Newport	2	180			180
Ross	26	534			2,993
Sligo	20	1,000			2,445
Strangford	29		1,406		1,406
Waterford	208	3,495			11,533

Wexford	61	1,960		2,166
Wicklow	51	245		1,424
Youghal	40	571		2,209
Total	4,688	43,516	210,522	278,125

From PRO. Estimate of tonnage shipping employed in trade of Ireland (1753) 15/54–15/56

APPENDIX 3

Charcoal-burning ironworks, 1600-1800

The following list is of ironworks known to have been in use at various times during the seventeenth and eighteenth centuries. The number in brackets refers to the Six Inch Ordnance Survey sheet on which the site can be located. The dates following a name show when the works was known to have been in production, but it does not follow that it was continuously in production throughout that period.

ANTRIM
Ardoyne (60), 1640. Belfast, end of eighteenth century. Carrickfergus (52), 1620. Lambeg (64), 1665. New Forge (65), 1630–41. Old Forge (65). Randalstown (43), 1642–55. Toome (42, 48), 1608. Whitehouse (57), seventeenth century.

CARLOW
Clonegall (18), 1737–42. Idrone (1), 1635—post 1641.

CAVAN
Arvagh (24, 30), 1734. Swanlinbar (7), 1700–39.

CLARE
Ballyrooaun (21, 29), 1690–1764. Feakle (20, 28), 1750. Furnace (21), 1700. Scarriff (28, 29), 1632–1750. Six Mile Bridge (52), 1650. Tomgraney (28), 1632. Whitegate (21a).

CORK
Adigole (103, 116). Araglyn (28), 1625–1770. Awenbeg (63). Bantry (118), 1684. Ballynetra (75), 1606–38. Bandon Bridge (110), 1642. Castle Martyr (77), 1764. Clonmeen (31), 1660–90. Coomhola (105), 1701. Cork city, 1740–97. Dunboy (115, 128), 1680. Dundaniel (96, 97), 1611–41. Fermoy (35). Glanmire (74), early eighteenth century. Glengarriff (90). Greenfield (28), mid-eighteenth century. Kilmacow (37, 46), 1606–22. Macroom (70, 71), early seventeenth century. Mallow (33), early seventeenth century. Mogeely (37, 46), 1593. Roaring Water (141), early eighteenth century. Youghal (67), 1607.

DONEGAL
Castlefinn (79), 1630? Letterkenny (53), 1752. Milford (36). Templecarn parish (95, 100, 101, 102, 104, 105). Stranorlar 78, mid-eighteenth century.

DUBLIN
Several foundries in the city, 1750–1800. Milltown (18, 22), 1737–86.

DOWN
Hill Hall (9, 15), 1625. Kilmore (30), 1630–54. Magheralin (13), 1667–73. Newry (46, 50), 1783.

FERMANAGH
Castlecauldwell (4, 9), 1611–41. Clonelly (5), 1611–1758. Drumcro (22, 27), 1615–41. Garrison (13), 1643.

GALWAY
Buffy Lough (67), 1754. Creggs (20), 1730–50. Eyrecourt (100, 108), 1741–80. Lough na furnace (53, 66), 1754. Martin's Mills (56), 1754. Screeb (52, 65). Woodford (125, 131) and Ballyrooaun (Clare, 21 29), 1690–1764. Woodroof (unidentified), 1741.

KERRY
Blackstones (72), c 1670–1754. Blackwater (92, 101), 1701–35. Brewster Field (67), 1696–1756. Dingle (43), 1600–1707. Dromore (92, 101), early eighteenth century. Drumkeane (89, 98). Glanerought (93), 1667–1754. Glencar (72), 1752. Glenmore (108), early eight-

eenth century. Gortalinny (93), 1689. Gortamullin (93), 1669-89. Muckross (66, 74), 1756. Tuosist (101), 1708.

KILDARE
Leixlip (11), 1753-93.

KILKENNY
Callen (26), 1734-59. Castlecomer (5), 1635-c 1770. Clohogue (5), c 1730.

LEITRIM
Ballinamore (25), 1695-1764. Creevelea (6), 1641-c 1768. Doubally (1), 1641. Drumod (35), 1695-1798. Drumshanbo (23), c 1640-1765. Drumsna (31), seventeenth and eighteenth century. Mullinalack (unidentified), 1740-45. Woodford (26), 1747-68.

LEIX
Ballyadams (19). Ballynakill (31), early seventeenth century—1652. Dysart (13, 18), late sixteenth century—1661. Killeskin (32, 37). Mountmellick (8) 1630-1756. Mountrath (16, 17), 1641-1792. Portarlington (4, 5), c 1660.

LIMERICK
Glin (17). Loghill (9, 18).

LONDONDERRY
Castledawson (42), late seventeenth century—1762. Draperstown (40), pre-1640—late eighteenth century. Drumconready (36), early seventeenth century. Dungiven (24, 25), 1726. Kilrea (27), late seventeenth century. Maghera (36). Salterstown (47, 49), 1626-54. Tullylinksay (42), late seventeenth century—1750.

LONGFORD
Cleenrath (3) and Enaghan (3), seventeenth century.

MAYO
Coolaght (91), 1741. Colounsolagh (unidentified), 1741. Foxford (60), early eighteenth century—1760. Islandeady (78). Knappagh (88, 98), 1687-93. Mullinmore (38), 1750-52. Port Royal (99), late seventeenth —late eighteenth century. Raheens (78), 1740. Tallaghan (9, 10), early eighteenth century.

MEATH
Clonard (41, 47). Sarney (50), early seventeenth century.

ROSCOMMON
Arigna (38), 1641–1788. Athleague (41). Ballyfarnan (1), seventeenth century. Boyle (5, 6), 1659–1763.

SLIGO
Ballynakill (27). Ballysadare (20). Kilmacteige (30, 36), eighteenth century. Screen (13, 19), 1768.

TIPPERARY
Cranagh (29, 30), 1730–65. Gortnahalla (34, 40). Old Forge (unidentified), 1742. Roscrea (12).

TYRONE
Derryvale (46, 47), 1690. Kirlish (24, 33), 1740. Lissan (29), early seventeenth century—1695.

WATERFORD
Ardmore (40). Cappoquin (21), 1615–1750. Dromslig (36). Lisfinny (28), 1620–85. Lismore (21), 1615–20. Minehead (29), 1591. Salterbridge (21), 1620. Tallow (28), 1588–1700.

WESTMEATH
Kilbeggan (38), c 1640.

WEXFORD
Camolin (11, 16). Coolgreany (3). Enniscorthy (20), 1560–1792. Forge (19), 1669–1764. Johnstown (4), 1763. Monart (19), 1660–1765.

WICKLOW
Aughrim (34). Avoca (40). Ballard (43), 1760. Ballycapple (30, 31). Ballinaclash (35), seventeenth century—1817. Carnew (47), 1640. Glenmalure (29). Glendalough (23), seventeenth century. Shillelagh (43), c 1641–1767. Vale of Clara (23). Wooden Bridge (40).

APPENDIX 4

Glossary of timber trade terms

Balk. Roughly squared large pieces of timber used for beams, etc, and measuring between 11 and 20in square.
Clapboards. Also called cleftboard or weatherboard. Fan-shaped pieces of wood used for roofing or shingling.
Compass and knee timber. Large pieces with a pronounced bend, which could be used in ship construction. Because of their shape the cost of transportation was high. The most suitable trees for this type of timber grew in hedgerows or in open woods where the branches had room to spread.
Deals. The term was first applied to softwood imported from the Baltic. While the size of deals varied it lay between certain limits: in the eighteenth century a standard deal was 12ft long, 11in wide, and 1½in thick, but Christiana deals were between 10 and 12ft long and 3in thick. A load of Danzig deals was 50 cubic ft.

Although the current nomenclature is red deal for Scots pine and white deal for Norway spruce, in the eighteenth century deals were referred to as either Norwegian or Spruce—a corruption of Prussia.
Hundreds. Timber exported from Danzig was counted by the

Shock of 60 pieces. Thus two Shock was 120 pieces. By the seventeenth century when the import of timber from eastern Europe was an established trade in Britain timber was being counted by the Long Hundred of 120 pieces, ie 2 Shock. Thus 200 deals actually meant 240 deals. This method of counting was also applied to staves. An Act of 1703 imposing a duty on staves specifically says 'the hundred of a hundred and twenty'.

Load. A load of timber was originally the weight carted by one horse and was about $1\frac{1}{4}$ tons. A load of plank was 50 cubic ft.

Plank. This was a more elastic term than deal and was applied to sawn wood between 1 and 8in thick. In customs returns, because of the variations in size, the value of plank was given and not the number of pieces.

Timber. This term denotes unsquared logs, or timber in the round.

Treenails. Thin cylindrical pieces of wood used to secure planking to a ship's timbers.

Bibliography

PRIMARY SOURCES

A. MANUSCRIPT

British Museum
Add MSS 4759. Customs returns, Ireland, 1682–86

HM Customs and Excise Library, London
Customs returns, Ireland, 1696, and various years for the eighteenth century

Public Record Office, London
15/54–15/56. Customs returns, Ireland, 1751, 1753, 1756, 1761.
Estimate of tonnage shipping employed in trade of Ireland, 1753
1/49. Minute book, no 49
SP 63/225. State Papers, Ireland

Armagh Museum
Brief survey of leases . . . manor of Brownlows Derry in the county of Armagh, 1 May 1667
Sam Johnston's valuation of the woods in the following townlands . . . 25 August 1748

Armagh Public Library
Henry Dillon's fine to Math Foster, Hillary 1657
John Dillon's settlement, 16 July 1731
Letters to Walter Harris, the Lodge Collection

Walter Harris, County Fermanagh, c 1740
Rev William Henry's topographical descriptions: a natural history of the parish of Killasher, 1 September 1732; account of county Fermanagh, 1739; account of county Antrim

Linenhall Library, Belfast
Letters to and from John O'Donovan during the process of the first Ordnance Survey. Typewritten MSS, ed O'Flanagan

National Library of Ireland
MSS 8110. An account of ships belonging to the port of Carrickfergus, 1676
MSS 8535 (11), (13), (15). Collection of Sir Richard Clayton
MSS 10934. Oliver estate and family papers
Ainsworth, J. Reports on MSS collections in Ireland in private hands.
Customs returns, 1764–93

Public Record Office, Ireland
Books of Survey and Distribution for counties Clare, Cork, Galway, Kildare, Kilkenny, Longford, Westmeath, Wexford
Hamilton MSS
Clayton MSS, 1A. 41. 41
Sarsfield-Vesey Correspondence

Public Record Office, Northern Ireland
T359. Archdale MSS
T473. Antrim MSS
T552. Calendar of the volume entitled 'Miscellaneous matters' in Armagh Registry by the Rev J. B. Leslie
T848. Ashe, T. View of the archbishopric of Armagh, 1703
T615. Harleian MSS
T904. A survey and valuation of the estate of the Rt. Hon. Alexander earl of Antrim, made in the year 1734 by Arch. Stewart, land surveyor.
T951. A brief survey of the several leases . . . within the manor of Brownlows Derry . . . 1667
T971. Brownlow papers
95A. The Abercorn papers
D580/265. Ely Estate leases
D552/215. Savage Estate papers

Royal Irish Academy
Ordnance Survey manuscripts, 1830s

B. PRINTED

Directories
Lucas. R. *Cork Directory* (1787)
Nixon. *Cork Almanach* (1797)
Watson, J. *The gentleman's and citizen's almanach* (1760, 1761, 1766)
West. *Cork Directory* (1810)
Wilson, P. *Dublin Directory* (1751, 1752, 1763-1800)

Documentary collections
Acts Privy Council (1615-26)
Calendar of Carew papers (1515-1624)
Calendar of state papers domestic (1667, 1670, 1671)
Calendar of state papers Ireland (1588-1670)
Commons' Journals, Ireland, 5, 8, 47, 48
Danvers, F. C. (ed). *Letters received by the East India Company*, 6 vols (London 1891)
Dunlop, R. *Ireland under the Commonwealth*, 2 vols (Manchester 1913)
Grosart, A. B. (ed). *The Lismore papers*, 8 vols (1880)
Henry, W. *Upper lough Erne in 1739*, ed G. S. King (Dublin 1892)
Hickson, M. A. (ed). *Selection of old Kerry records* (1872)
Hill, G. (ed). *The Montgomery manuscripts, 1603-1706* (Belfast 1869)
Hogan, E. (ed). *The history of the war in Ireland from 1642 to 1643* (Dublin 1873). *The description of Ireland and the state thereof 1598* (Dublin 1878)
Longfield, A. K. (ed). *The Shapland-Carew papers* (Dublin 1946)
MacLysaght, E. (ed). *The Kenmare manuscripts* (Dublin 1942)
Minutes of the Dublin Society, 1768-1800
Phillips, T. *Londonderry and the London Companies*, ed D. A. Chart (Belfast 1928)
Sainsbury, E. B. (ed). *Calendar of the Court Minutes of the East India Company*, 2 vols (1907, 1909)
Simington, R. C. (ed). *The Civil Survey*, 10 vols (Dublin 1931-53). *Books of Survey and Distribution*, vol 1, Roscommon (Dublin 1949)
Statutes at Large

Newspapers
Belfast Newsletter (1750, 1753-60, 1762-7)
Dublin Intelligence (1728-9)
Dublin Weekly Journal (1725-6, 1730, 1747-52)

Faulkner's *Dublin Journal* (1730–1800)
Freeman's *Journal* (1763–8, 1770–1)
Limerick Chronicle (1769, 1771–2, 1774, 1777)
Munster Journal (1749–50, 1766, 1777)
Pue's Occurrences (1736, 1740, 1746–7)
Universal Advertiser (1753)

C. THESES

McCracken, E. M. 'The composition and distribution of woods in Northern Ireland from the sixteenth century down to the establishment of the first Ordnance Survey', unpublished MSc thesis in the library of the Queen's University, Belfast (1944)

McEvoy, T. 'Some Irish native woodlands: an ecological study', unpublished M Agi Sc thesis in the library of National University Dublin (1945)

D. MAPS

BM
C.2.cc 1.	Map of Ireland, by B. Boazia, c. 1600
C.3.c.5.	Irlandiae, by Mercator
Maps C.2.	The counties of Leinster, Mounster, Ulster and Connaugh, by J. Speede, 1610
9 Tab 21.	The provinces of Leinster, Mounster, Ulster and Connaugh, by J. Jansson, 1647
Maps C.4.c.1.	Leinster, Mouster, Ulster and Connaughir, by W. Blaeu, 1662
K51 (2).	Hiberniae, Britannicae Insulae, nova descriptio by N. Sanson, 1665
Maps 19 d.1.	The provinces of Leinster, Munster, Ulster and Connought, by W. Petty, 1690

PRO
MPF 67.	Plott of Ireland by J. Norden from Boazio, c. 1610
MPF 70.	The description of the barony of Idrone, c. 1580
MPF 71.	Sligo and Mayo by J. Browne and J. Baptiste, 1587
MPF 54.	Munster, c. 1580
MPF 76.	County Monaghan, by J. Browne and J. Baptiste, 1590?
MPF 77.	Belfast lough, 1590
MPF 87.	Map of the coast of Ireland from Knockfergus to Dundrum, c. 1580

MPF 92.   Mayo, by J. Baptiste, 1585
MPF 97.   Limerick by F. Jobson, 1587
MPF 100.   Munster, Kinsale to Dingle, 1587?

SECONDARY SOURCES

Albion, R. G. *Forests and sea power. The timber problem of the Royal Navy, 1652-1862* (1926), Cambridge, Harvard
Analecta Hibernia, nos 6, 8
Anderson, M. L. *The selection of tree species*, (1961). 'Items of forestry interest from the Irish statutes prior to 1800', *Irish Forestry* (1943), 1, no 2
Andrews, J. H. 'Notes on the historical geography of the Irish iron industry', *Irish Geography*, iii no 3
Archer, J. *Statistical survey of county Dublin* (Dublin 1801)
Atkinson, A. *Ireland exhibited to England*, 2 vols (Dublin 1823)
Atkinson, E. D. *An Ulster parish . . . Donaghcloney* (Dublin 1898)
Bagenal, H. 'The description and present state of Ulster', *UJA*, 2nd ser, ii no 4
Bagwell, R. *Ireland under the Stuarts* (1909)
Bale, M. P. *A book of sawmill and wood converting machinery* (1919)
Barrow, J. *A tour round Ireland* (1836)
Beaufort, D. A. *Memoir of a map of Ireland* (1792)
Beaufort, W. 'The ancient topography of Ireland', in *Collectanea de rebus Hibernicus*, iii, Charles Vallencey (Dublin 1782)
Belmore, Earl of. 'Ancient maps of Enniskillen', *UJA*, 2nd ser, ii no 4. *History of two Ulster manors* (Dublin 1903)
Benn, G. *History of Belfast* (1877)
Bloome, R. *A description of Ireland* (1673)
Boate, G., Molyneux and others, *A natural history of Ireland* (Dublin 1755)
Boutcher, W. *Treatise on forest trees* (Dublin 1784)
Brereton, W. *Travels in . . . Ireland, 1634-5*, ed E. Hawkins (Manchester 1844)
Bush, J. *Hibernia curiosa* (1769)
Carr, J. *The Stranger in Ireland* (1806)
Caulfield, R. *The Council Book of the Corporation of Youghal* (Guildford 1878)
Carter, B. Letter in *The Times*, 3 August 1963
Chambers, E. *Cyclopaedia* (1738)
Chart, D. A. 'The breakup of the estate of Con O'Neill, Castlereagh, county Down, temp. James 1', *PRIA*, xlviii, c3

Clear, T. 'Forestry', *The natural resources of Ireland* (Dublin, 1944), 81–90
Considerations concerning balance of trade between English and foreign iron (1661)
Concise view of the origin . . . of the Irish Society (1822)
Coote, C. *Statistical survey of county Armagh* (Dublin 1804)
— *Statistical survey of the King's county* (Dublin 1802)
— *Statistical survey of county Monaghan* (Dublin 1801)
Dawson, A. 'Notes on re-afforestating of Ireland from the parochial records of Seagoe parish', *RHASIJ*, 4th ser, vi, (1883–4)
Dobbs, A. *Observations of the trade and improvements of Ireland* (Dublin 1729)
Dowdall, N. 'A description of county Longford, 1682', *Ardagh CASJ*, i no 3
Downings, R. 'Phillipps MSS, county Longford', ibid, i no 3
Dublin Penny Journal (1832–4)
Dublin Society Weekly Observations (Glasgow 1756)
Dutton, H. *Statistical survey of county Clare* (Dublin 1824)
— *Statistical survey of county Galway* (Dublin 1824)
Edlin, H. L. 'Review of *The forests of Ireland*', *Quarterly Journal of Forestry*, lxi no 2 (1967), 176
Elwes and Henry, *The trees of Great Britain and Ireland* (Edinburgh 1906–13)
Erdtman, G. 'Traces of the history of the forests of Ireland', *Irish Naturalists Journal*, i no 12
Falkner, C. L. *Illustrations of Irish history* (1904)
— 'The forest question considered historically', *Statistical and Social Society of Ireland* (23 Jan 1903, Dublin 1903)
Fincham, J. *A history of naval architecture* (1851)
Fitzpatrick, H. M. 'The trees of Ireland, native and introduced', *PRIA*, 20, NS no 41
— *The forests of Ireland* (editor) (Bray 1966)
Forbes, A. C. 'Some legendary and historical references to Irish woods and their significance', *PRIA*, xli B3
— 'Tree planting in Ireland during four centuries', ibid, xli C6
— 'Some early economic and other developments in Eire, and their effect on forestry conditions', *Irish Forestry*, i no 1 (1943), 6
— 'The forestry revival in Eire', ibid, iv no 1 (1947)
Frazer, R. *A general view of the county Wicklow* (Dublin 1901)
Freeman, T. W. 'Forestry and Land Use Survey', *Irish Forestry* (1950) vii, 24
Frost, J. *History of county Clare* (Dublin 1893)

Graham, H. G. *Social life of Scotland in the eighteenth century* (1937)
Halley, E. *Atlas maritimus et commerciales* (1728)
Hammersley, G. 'The crown woods and their exploitation in the sixteenth and seventeenth centuries', *Bulletin Institute Historical Research*, xxx (1957)
Harris, J. *Lexicon Technicum* (1710)
Harris, W. *Remarks on the affairs and trade of England and Ireland* (1691)
Harris, W. *Hibernica: or some ancient pieces relating to Ireland* (Dublin 1770)
— *The ancient and present state of the county Down* (Dublin 1744)
Hart, C. E. *Royal Forest, a history of Dean's woods as producers of timber* (Oxford 1966)
Hayes, R. *The negociator's magazine* (1749)
Hayes, S. *A practical treatise on planting* (Dublin 1822)
Hennessy, P. *Raleigh in Ireland* (1883)
Henry, A. 'Woods and trees in Ireland', *Louth AJ*, iii no 3
— 'Irish forestry', *Cumann Leigheacht an Phobail*, ser B, no 2
Hill, G. *The Stewarts of Ballintoy* (Coleraine 1865)
'Historical and topographical notes', *Cork HASJ*, 2nd ser, xxi
Hodgson, L. *The complete measurer* (Dublin 1801)
Hodson, R. E. 'Woodland of west Cork two hundred years ago', *Cork HASJ*, 2nd ser, viii
Holtzapffel, C. *Turning and mechanical manipulation*, 2 vols (1843)
Hore, H. F. 'Particulars relative to Wexford and the barony of Forth by Colonel Soloman Richards, 1682', *Kil ASJ*, new ser, iv, pt 1
— 'Woods and fastnesses in ancient Ireland', *UJA*, 1st ser, vi
Jones, P. and Simmons, E. N. *Story of the saw* (1961)
Joyce, P. W. *Irish local names explained* (Dublin undated)
Kearney-Jones, A country gentleman. *Essays on agriculture and planting founded on experiments made in Ireland* (Dublin 1790)
Kelly, P. *Universal Cambist* (1821)
K'Eogh, J. *Botanologia universalie Hibernica* (Dublin 1735)
Kilpatrick, C. 'Public response to forest recreation in Northern Ireland', *Irish Forestry*, xxii no 1 (1965), 3
Laslett, T. *Timber and timber trees* (1894)
Le Fanu, V. C. 'The royal forest of Glencree', *RSAIJ*, xii, pt 3
'Letter from Portadown from W. Brooke to W. Molyneux, 1682', *UJA*, 2nd ser, iv
Lewis, S. *Topographical dictionary of Ireland*, 2 vols (1837)
Longfield, A. *Anglo-Irish trade in the sixteenth century* (1929)
Loudon, J. C. *Arboretum et fruticetum Britannicum* (1844)

Lucas, A. T. 'The dugout canoe in Ireland. The literary evidence', *Sartruck ur Varberge museum* (1963)
— 'The sacred trees of Ireland', *Cork HASJ*, lxviii (1963)
McCracken, E. M. 'Charcoal burning ironworks in seventeenth and eighteenth century Ireland', *UJA*, xx (1957)
— 'Supplementary list of Irish charcoal burning ironworks', ibid, xxviii (1965)
— 'The woodlands of Ulster in the early seventeenth century', ibid, x (1947)
— 'The woodlands of Donegal', *Donegal Annual*, iv no 1
— 'The woodlands of Ireland c. 1600', *Irish Historical Studies*, xi no 44
— 'The Londonderry woodlands', *County Londonderry Handbook* (Belfast 1964)
— 'Irish woodlands, 1600 to 1800', *Quarterly Journal of Forestry*, lvii no 2
— 'Irish nurserymen and seedsmen, 1740 to 1800', ibid, lix no 2
— 'The Irish timber trade in the eighteenth century', ibid, lxi no 1
— 'The Irish timber trade in the seventeenth century', *Irish Forestry*, xxi no 1
— 'Notes on eighteenth century Irish nurserymen', ibid, xxiv no 1
M'Evoy, J. *Statistical survey of county Tyrone* (Dublin 1802)
McEvoy, T. 'Irish native woodlands—their present condition', *Irish Forestry*, i, pt 2 (1943)
McGrath, P. *Merchants and merchandise in seventeenth century Bristol* (Bristol 1955)
Mackay, J. *Forestry in Ireland* (Cork 1934)
Mackay, J. T. *Flora Hibernica* (Dublin 1836)
MacLysaght, E. *Irish life in the seventeenth century* (Cork 1950)
McNamara, M. 'Extracts from "Report on re-afforestation of Ireland"', *Irish Forestry*, xxiv, pt 2 (1967)
McParlan, J. *Statistical survey of Donegal* (Dublin 1802)
— *Statistical survey of Leitrim* (Dublin 1802)
— *Statistical survey of Sligo* (Dublin 1802)
Mallenson and Leigh, *Timber trade practice*
Marshall, J. J. *History of Charlemont fort* (Dungannon 1921)
Mason, W. *Statistical and parochial survey of Ireland*, 3 vols (Dublin 1814)
Memoirs Geological Survey to various sheets
Meyer, K. *Selections from ancient Irish poetry* (1911)
Miller, J. R. *Speeches in the House of Commons upon the equalization of the weights and measures of Gt. Britain* (1790)

Mitchell, G. F. 'Littleton Bog, Tipperary: an Irish vegetational record', *Geological Society America Institute*, Special Paper, 84 (1965)
Moody, T. W. *The Londonderry Plantation, 1609–41* (Belfast 1939)
— 'Redmond O'Hanlon', *Proc Belfast Natural History and Philosophical Society*, 2nd ser, 1, pt 1
More, A. G., Colgan, N. G. and Scully, R. W. *Contributions towards a Cybele Hibernice* (Dublin 1898)
Murphy, G. *Early Irish lyrics* (Oxford 1956)
Murray, L. P. 'Irish diary of the confederate wars', *Louth AJ*, vii no 1
Newenham, T. *A view of the natural circumstances of Ireland* (1809)
O'Brien, G. *Economic history of Ireland in the seventeenth century* (Dublin 1919)
— *Economic history of Ireland in the eighteenth century* (Dublin 1918)
O'Donovan, J. *Economic history of livestock in Ireland* (Cork 1940)
O'Grady, H. *Strafford and Ireland*, 2 vols (Dublin 1923)
O'Longain, S. article in *An claidheamh Soluis*, x no 10, 7
O'Sullivan, W. *Economic history of Cork city* (Cork 1937)
Otway, C. *Sketches in Ireland* (Dublin 1827)
Parliamentary Gazetteer of Ireland (Dublin 1846)
Petty, W. *The political anatomy of Ireland*, 1672 (1691)
Pinkerton, W. 'Contributions towards a history of Irish commerce', *UJA*, 1st ser, iii
Porritt, E. and A. G. *The unreformed House of Commons* (1903)
Power, P. 'Battle of the Comeraghs mountains, 1643', *Waterford Archaeological Society Journal*, lx
Prendergast, J. P. 'Extracts from the Journal of Thomas Dineley', *Kilkenny ASJ*, new ser, i, pt 1
— 'The plantation of Idrone', ibid, new ser, iii, pt 1
— 'The Tory war of Ulster', ibid, new ser, vi
Querist, (The) (Dublin 1725)
Radcliff, T. *Report of agriculture of county Kerry* (Dublin 1814)
Report from the select committee on industries (Ireland) (1885)
Ryan, J. *History of the county Carlow* (Dublin 1833)
Sampson, G. V. *Statistical survey of county Londonderry* (Dublin 1802)
Schubert, H. P. *History of the British iron and steel industry from c 450 B.C. to 1775* (1957)
Seward, W. W. *Topographia Hibernica* (Dublin 1795)

Sharkey, M. 'The timber industry in Ireland', paper given to the Institute of Civil Engineers of Ireland (1966)
'Shawn Ru, the rapparee', *Cork HASJ*, 2nd ser, xi
Sheffield, Lord John. *Observations on the manufactures of Ireland* (1785)
Simms, J. G. 'The Civil Survey, 1654–6', *Irish Historical Studies*, ix no 35
— *The Jacobite parliament* (Dublin 1965)
Smith, C. *Ancient and present state of Cork*, 1749, 2 vols (Cork 1815)
— *Ancient and present state of Kerry*, 1756 (Dublin 1774)
— *Ancient and present state of Waterford* (Dublin 1746)
Stebbing, W. *Sir Walter Raleigh* (Oxford 1891)
Stevenson, J. *Two centuries of life in Down, 1600 to 1800* (Belfast 1920)
Tansley, A. G. *The British Isles and their vegetation*, 2 vols (Cambridge 1953)
Thompson, R. *Statistical survey of county Meath* (Dublin 1802)
Threlkeld, C. *Synopsis Stirpium Hibernicanum* (Dublin 1726)
Tighe, W. *Statistical observations on county Kilkenny* (Dublin 1802)
Townsend, H. *Statistical survey of county Cork* (Dublin 1815)
Wakefield, E. *An account of Ireland*, 2 vols (1812)
Wallace, T. *Essay on the manufactures of Ireland* (Dublin 1798)
Warburton, J. *History of the city of Dublin*, 2 vols (1818)
Weld, I. *Illustrations of the scenery of Killarney* (1807)
— *Statistical survey of county Roscommon* (Dublin 1832)
Westropp, T. J. *Irish glass* (1920)
— 'Carrigogunnell Castle', *RSAIJ*, xxxvii pt 4
— 'Ring forts in Moyarta barony, county Clare', ibid, xxxix pt 2
Wood-Martin, W. G. *History of Sligo, 1603–88* (Dublin 1889)
Wright, H. L. 'Our Journal', *Quarterly Journal of Forestry*, lx, pt 1 (1966)
Young, A. *Tour of Ireland 1776–9*, ed Hutton, A. W., 2 vols (1892)
Young, R. M. *Historical notes of old Belfast* (Belfast 1896)

Index

Italic numerals refer to illustrations

Acreages, 134, 140, 141, 146, 147
Advertisements, 18, 19, 63, 72, 118, 126, 132, 133, 134, 158
Aherlow wood, 37, 39, 45
Alder, 17, 20, 22, 39, 40, 42, 47, 54
Amenity, 104, 105, 149–50
Anderson, Mark Loudon, 143, 146
Apple, 26
Arbutus, 18, 46, 56, 95
Ash, 14, 17, 20, 23, 39, 40, 42, 47, 49, 60, 82, 88, 117, 124, 131, 136, 140, 147
Avondale, 86, 143

Balk, 114–16, 118, 169
Bark, 45, 48, 51, 57, 79–82
Barrel staves, 49, 59, 60, 98, 107, 124, *see also* pipestaves, hogshead staves and staves
Beech, 17, 18, 39, 42, 60, 117, 135, 136, 140, 147, 152
Birch, 17, 20, 22, 40, 42, 43, 46, 47, 49, 54, 80, 82, 136, 140, 147
Bird cherry, 17, 26
Blackthorn, 23, 26
Bog timber, 21, 60, 78
Boyle, Richard, Earl of Cork, 45, 46, 48, 80, 88, 89, 94, 100, 101, 126
Brewers and distillers, 60, 119, 120
Bridges, 86

Brousdon, Peter, 50, 99, 102, 123
Burren, 19

Cabins, 44, 78
Capercaille, 17
Cattle, 21, 42, 43, 48, 56, 78, 80, 109, 132
Cedar, 20, 136, 140, 152
Chamney family, 50, 95
Charcoal, 58, 59, 92
Civil Survey, 49, 161
Clanawle, 37, 40
Clanbrassel, 37
Clancan, 37, 39, 40
Clapboard, 98, 115, 116, 125, 169
Clapholt, 115, 116, 121
Clonish, 44, 55
Coaches, 116, 131
Cocoa wood, 117
Coillaughrim (wood), 37, 49
Coillconchobhair (wood), 44
Coopering, 59–62
Coopers, 57, 60–2, 126, 127
Coote family, 93, 94, 95
Coppice, 50, 81, 92, 95
Cots, 51, 62–3, 66, 100

Dargle glen, 56, 85
Deal, 77–8, 110
Deals, 58, 111–14, 118, 125, 130, 169

Deer, 13, 42, 49, 89
Derry, 23, 24–5, 26
Dublin Society, 81, 82, 87, 136, 138, 140
Dufferin, 39, 40

East India Company, 45, 46, 71, 101
Ebony, 117, 124
Elm, 17, 18, 20, 23, 39, 60, 65, 71, 75, 136, 140
Enniscorthy, 90, 92, 93, 94, 123, 126
Estates, 30, 31, 32, 80, 83, 101, 122, 126, 135, 138, 142, 146, 152
Export of timber, 98, 100–2, 105–13

Fasach Coille (wood), 37
Fen, 37, 39, 40, 42
Fencing, 136
Fens, 37, 44
Fire, 21, 27
Fireboot, 44, 78
Firs, 18, 26, 141, 147
Forbes, A. C., 143
Forest Parks, 103, 149–50
Forestry workers, *see* timber workers

Gladstone's Land Acts, 135, 141, 142
Glanekinty, 37, 46
Glass, 58, 87–90
Glengarriff, 18, 56, 166
Glenconkeyne, 17, 37, 39, 40, 54, 72, 105–6
Glens of Antrim, 18, 22, 26, 35, 39, 55, 56, 83
Goats, 21, 137
Grand juries, 32, 137
Great Belan, 34
Griffith, Sir Arthur, 143

Hawthorn, 20, 26, 136
Hayes, Samuel, 86, 143
Hazel, 17, 19, 23, 26, 39, 40, 42, 43, 44, 46, 54, 59, 121
Headings, 101, 102
Hedgerows, 20, 151, 152
Hides, 57, 81, 87
Hodgson, Levi, 129–30

Hogshead staves, 59, 60, 98, 102, 106, 107
Holly, 14, 17, 20, 22, 39, 40, 42, 46, 47, 56
Hoops, 59, 115, 116, 121
Horse chestnut, 17, 20, 39
Houses, 49, 73, 74, 75, 76, 77–9, 117, 118
Howitz, Daniel, 143, 150, 160
Hundreds, 169

Idrone, 33
Imports of timber, 110–21
Irish Society, 31, 105
Ironworks, 43, 44, 45, 46, 48, 50, 53, 54, 55, 57, 90, 91, 92–6, 99, 137, 165–8

Juniper, 17, 19, 20

K'Eogh, J., 19
Kilconish, Great Wood of, 37, 39, 49
Killarney, 17, 37, 39, 41, 45, 52, 54, 80, 99
Killetra, 37, 39, 40, 72, 75, 105, 106
Killultagh, 35, 39, 40
Kilmore, Great Wood of, 37, 39, 48
Kilwarlin, 17, 39, 40
Kirchoffer & Graves, 130
Knee and compass timber, 65, 123, 130, 169

Landlords, 54, 135, 141
Landscape planning, 149–52
Larch, 136, 140, 147
Leases, 30, 31, 54, 136, 137
Leguy woods, 37, 39, 42, 55
Lime, 17, 39, 135, 152
Littleton bog, 19
London companies, 31
Longe, George, 88
Lough Gill, 19, 42
Lucas, A. C., 62

McCartan's Country, 40
Mahogany, 79, 116, 117, 124, 127, 130
Mantraps, 138
Maple, 136, 140
Masts, 66, 71, 72, 99, 112, 116
Meeting of the Waters, 52

Index

Mountain ash, 17, 26, 40
Mountrath, 55, 92, 93, 94, 167
Mountreivelin, 37, 39, 40

National Forest Parks, 149–50
Navy, 46, 48, 50, 64, 123, 156
New Forest, 17, 40
Norway spruce, 147, 148
Nurseries, 138, 139, 140

Oak, 17, 18, 19, 20, 22, 23, 24, 25, 26, 39, 40, 42, 43, 44, 46, 47, 49, 53, 60, 68, 69, 73, 75, 77, 78, 79, 82, 95, 117, 122, 123, 134, 147; species of, 130, 136, 140
Oars, 71, 72, 116
O'Neilland, 37, 39
Orchards, 55

Parliamentary acts, 27, 32, 64, 82, 110, 121, 136–7, 149
Passes, 27
Petty, Sir William, 18, 45, 46, 50, 54, 66, 83, 93, 95
Pine, 17, 18, 20, 26, 65; species of, 136, 140
Pine martin, 22
Pinus contorta, 141, 147, 149
Pipestaves, 48, 49, 59, 60, 98, 101, 102, 105, 106, 107, 124
Plank, 98, 106, 107, 108, 109, 111, 116, 123
Plantations, 34, 103–4, 140–1
Poplar, 17, 18, 82, 136, 140
Ports, 99, 100, 101, 102, 105, 106, 108, 109, 117–18, 123, 130, 163–4
Provision trade, 59, 60, 61, 81, 96, 98, 109, 110, 119, 120
Prices, 80, 116, 117, 120–6, 131–2, 140, 149
Pounden family, 92

Rabbits, 21
Raleigh, Sir Walter, 45, 94, 99, 100
Rawdon, Sir George, 72, 73, 89, 99, 111
Royal Oak of Portmore, 73, 80
Rutledge, 55, 92

Sallow, 22
Sawing and sawmills, 119, 123, 124, 125
Scenic drives, 149
Scots pine, 17, 136, 138, 140, 147–8
Scrub, 30, 40, 43, 44, 49, 53, 54, 55, 56, 83, 161–2
Seasoning, 65, 111, 131
Settlers, 27, 28, 40, 47, 57, 58, 73, 78, 93, 98, 99
Shillelagh, 37, 50, 54, 64, 94, 100, 101, 102, 122, 123, 168
Shipbuilding, 46, 48, 58, 62–73, 101, 102
Shipyards, 34, 66, 67, 69, 70, 85
Sitka spruce, 103, 141, 146, 147, 148
Slieve Groot wood, 37
Spars, 71, 72, 112, 116
Staves, 56, 65, 98, 99, 100, 101–2, 105–11, 114–16, 120, 121, 124; see also pipestaves, hogshead staves and barrel staves
Suidain wood, 37, 43
Sweet chestnut, 60, 136, 140
Sycamore, 17, 18, 20, 39, 60, 135, 136, 147, 152

Tanneries, 45, 46, 48, 80–4, 87
Tanners, 57, 79–84, 87
Tenants, 32, 42, 77, 136, 137, 141
Thorn, 54
Threlkeld C., 18, 19, 21
Timber measurers, 127, 129–30
Timber merchants, 61, 126, 127, 128, 129–32, 134
Timber production, modern, 147; cost of, 149
Timber, sale of, 130–2
Timber workers, 99–100, 126, 148
Tollymore Forest Park, 150
Tory, 28, 50
Townlands, 22–6
Traditions, 21, 22, 39, 43, 47
Transport, 54, 72, 99, 105, 118, 119, 123, 124, 131
Tulip tree, 136

Wages, 58, 70, 93; see also prices
Wainscot, 73, 79, 98, 111, 117
Wakefield, E., 56, 84

Walnut, 111, 116, 124, 135, 136
Water pipes, 18
Walworth wood, 143
Weymouth pine, 136, 140
Whitebeam, 17
White family, 45, 54
Whitethorn, 23
Wild cat, 21, 43
Willow, 17, 39, 40, 59, 82, 121
Wolf, 13, 21, 28–30, 47

Wolfhound, 29, 30
Woodcock, 45
Wooden ware, 115, 116
Woodkerne, 27–9, 40, 47, 49
Wood rangers, 50, 131, 138

Yew, 17, 19–20, 23, 26, 42, 56, 78, 140
Young, Arthur, 49, 56, 118

DATE DUE

Demco, Inc. 38-293